Strengthening Geography In The Social Studies

Salvatore J. Natoli

Editor

National Council for the Social Studies
Bulletin No. 81

NCSSPublications

National Council for the Social Studies

Revised Edition Editor: Salvatore J. Natoli
Revised Edition Art Director: Gene Cowan

Library of Congress Catalog Card Number: 88-61299
ISBN 0-87986-056-1

Printed in the United States of America
10 9 8 7 6 5 4 3

Third Printing (Revised) 1994

TABLE OF CONTENTS

CONTRIBUTORS   v

PREFACE   ix
Salvatore J. Natoli

Chapter 1: MODERN GEOGRAPHY   1
Salvatore J. Natoli and Charles F. Gritzner

Chapter 2: THE STATUS OF GEOGRAPHY IN
           MIDDLE/JUNIOR AND SENIOR HIGH SCHOOLS   11
Joseph M. Cirrincione and Richard T. Farrell

Chapter 3: GEOGRAPHY WITHIN THE SOCIAL STUDIES
           CURRICULUM   22
Michael Libbee and Joseph Stoltman

Chapter 4: GEOGRAPHY IN THE SOCIAL STUDIES SCOPE
           AND SEQUENCE   42
Joseph Stoltman and Michael Libbee

Chapter 5: THE PREPARATION OF GEOGRAPHY TEACHERS   51
Dennis L. Spetz

Chapter 6: THE FUNDAMENTAL THEMES IN PRACTICE:
           Nuclear Explosion at Chernobyl—An Example of
           Global Interdependence   59
Robert W. Morrill, James Sellers, and Stephen A. Justham

Chapter 7: THE FUNDAMENTAL SKILLS OF GEOGRAPHY   72
George Vuicich, Joseph Stoltman, and Richard G. Boehm

Chapter 8: GETTING GEOGRAPHY INTO THE
           CURRICULUM   93
Part 1: Pursuing a Decalogue: James F. Marran
Part 2: Implementing a Geography Program: Salvatore J. Natoli
Part 3: Resources for Keeping Current with Geography: Joan Juliette

Chapter 9: TEACHERS ASSESS THE FIVE FUNDAMENTAL
           THEMES OF GEOGRAPHY   119
Richard T. Farrell and Joseph M. Cirrincione

INDEX   123

For
Charles T. Natoli, 1940–1986, and Fredric A. Ritter, 1932–1986,
Best friends, artful teachers, and enduring inspirations

CONTRIBUTORS

Richard G. Boehm is a professor of geography and chair of the Department of Geography and Planning at Southwest Texas State University in San Marcos. In 1983, he served as president of the National Council for Geographic Education (NCGE). He was a contributing author to *Guidelines for Geographic Education*, published in 1984. His B.S. (education) and M.A. (geography) are from the University of Missouri, Columbia; his Ph.D. (geography) is from the University of Texas. He has had a lifelong interest in improving geographic education and is currently co-coordinator of the National Geographic Society (NGS) sponsored Texas Alliance for Geographic Education. He also served on the Geographic Education National Implementation Project (GENIP) steering committee, 1985–1987, and has many publications, including articles in *Social Education* and the *Journal of Geography*.

Joseph M. Cirrincione is an associate professor at the University of Maryland at College Park and holds a joint appointment in the Geography Department and the Department of Curriculum and Instruction. A former junior high school teacher in Brooklyn, New York, he currently teaches undergraduate and graduate courses in geography and social studies education and supervises student teachers. His B.S. (education) is from the State University College of New York at Oswego; he holds an M.A. (geographic education) and a Ph.D. (geographic education) from the Ohio State University. His recent publications have focused on teaching geography and history at the upper school level.

Richard T. Farrell is an associate professor at the University of Maryland at College Park where he holds a joint appointment in the Department of History and the Department of Curriculum and Instruction. His teaching experience includes work at the junior and senior high school levels. He currently teaches graduate and undergraduate courses in American history and social studies education and supervises student teachers. His recent publications have appeared in history journals, as well as in *Social Education, Journal of Reading,* and *Theory and Research in Social Education*. Recent presentations at national and regional conferences have included work on teaching geography and history at the upper school level. He holds a B.A. (history) from Wabash College, and M.S. (secondary education) and Ph.D. degrees (history) from Indiana University.

Charles F. Gritzner is a professor of geography at South Dakota State University. He is a former executive director and president of the National Council for Geographic Education. He has taught at the college and university level for more than twenty years at Louisiana State University, University of Houston, East Carolina University, and Western Oregon College. He has more than 100 publications on geographic education and on cultural geography. He holds a B.A. (geography) from Arizona State University, and an M.A. and Ph.D. in geography from Louisiana State University. Dr. Gritzner has made hundreds of workshop and professional meeting presentations on teaching geography and cultural geography.

Joan Juliette is an elementary school librarian who will complete a Master of Arts degree in geography in 1988 at Indiana University of Pennsylvania. As a geography graduate student, she began compiling lists of resources for classroom teachers of geography. She received a B.S. degree in library science from Clarion University of Pennsylvania and holds a master's degree in Library Science from Indiana University in Bloomington. She has identified children's books and computer programs for geography and has integrated them into the numerous weekly library lessons for which she is responsible in the Armstrong School District, Elderton, Pennsylvania.

Stephen A. Justham is an associate professor and chair of the Department of Geography at Kutztown University of Pennsylvania. He received a B.S. and M.A. in geography from Indiana University of Pennsylvania and a Ph.D. (geography) from the University of Illinois. He has had teaching experience at the secondary school level. He has conducted research on wind energy, dendroclimatology, and adult education. He has coauthored four books and has published articles in professional journals. He is currently serving as Center Director for a Fund for the Improvement of Postsecondary Education grant for setting up centers for Excellence in Geographic Education.

Michael Libbee is an associate professor of geography at Central Michigan University, Mount Pleasant, Michigan. His primary teaching responsibilities are with preservice teachers. His publications include articles on teaching techniques appropriate to both the secondary school and college levels. He holds a B.A. (geography) and an M.A. and Ph.D. (geography) from Syracuse University. He has just concluded a project funded by the Exxon Education Foundation and the Johnson Foundation on fostering international understanding through world geography.

James F. Marran is a geography teacher and chair of the Social Studies Department at New Trier Township High School in Winnetka, Illinois. He was one of the original trial teachers with the High School Geography Project (HSGP) and has conducted many workshops around the country for the project. He also served on the Steering Committee for the HSGP. He is chair of the Steering Committee for the Geographic Education National Implementation Project (GENIP). He holds a B.A. (history) from Holy Cross and an M.A. (history) from the University of Maryland. He has published articles and text materials on geography themes and topics in history and the social sciences. His present interests are education reform movements at the state and national levels.

Robert W. Morrill is an associate professor and head of the Department of Geography at the Virginia Polytechnic Institute and State University. He has been active in geographic education and social studies education since beginning his teaching career in Massachusetts as a junior high school social studies teacher. He was a contributing author to *Guidelines for Geographic Education,* published in 1984, and is a member of the Steering Committee of GENIP. He received a B.A. (political science) and M.A. (social science) from Assumption College and a Ph.D. (geography) from Clark University, both in Worcester, Massachusetts. His publications have dealt with the improvement of geographic education. He is first vice president and president-elect of the National Council for Geographic Education.

Salvatore J. Natoli is former Director of Publications for the National Council for the Social Studies and editor of *Social Education.* He was formerly Deputy Executive Director of the Association of American Geographers (AAG) and Project Director of GENIP. He has taught geography and social studies in junior and senior high school. He taught courses in physical and cultural geography at Mansfield University of Pennsylvania, Clark University, and the University of Connecticut. He directed Title XI NDEA institute and fellowship programs in geography and the social sciences with the U.S. Office of Education. He was a staff geographer on the HSGP. He chaired the joint committee of AAG and NCGE that developed *Guidelines for Geographic Education.* His B.S. (geography and social sciences) is from Kutztown University of Pennsylvania and his A.M. and Ph.D. (geography) are from Clark University, Worcester, Massachusetts. He is author or editor of more than sixty publications on various subfields of geography and on geographic education. His most recent publication is *Geography in Internationalizing the Undergraduate Curriculum.*

James L. Sellers was graduated from Virginia Polytechnic Institute and State University with a B.A. in geography. After teaching high school geography for ten years in the Montgomery County, Virginia, Public Schools, he became Supervisor of Curriculum for that school system. He received an M.A. (social studies education) and Ph.D. (curriculum and instruction) from Virginia Polytechnic Institute. He worked extensively with the Virginia Department of Education on the 9th grade social studies Standards of Learning program and on state geography textbook adoptions. He is principal of Price's Fork Elementary School in Blacksburg, Virginia, and is active in national and state social studies and elementary principal's organizations. He is on the executive board of the Virginia Council for the Social Studies.

Dennis L. Spetz is associate professor and distinguished teaching professor in the Department of Geography at the University of Louisville, Kentucky. He received an A.B. (geography) from Harpur College (SUNY Binghamton), an M.A. (geography) from Kent State University, and an Ed.D. from Indiana University. He has served as university ombudsman and associate dean of the College of Arts and Sciences. In 1984, he was appointed first state geographer in Kentucky. He is chair of the GENIP Teacher Certification Committee and coordinator of the Kentucky Geographic Alliance.

Joseph P. Stoltman received a B.A. (geography) from Central Washington University, an M.A. (geography) from the University of Chicago, and an Ed.D. (social studies education) from the University of Georgia. He has taught geography, social studies, and language arts at all levels of the elementary and secondary schools. He is a professor of geography at Western Michigan University, Kalamazoo, Michigan. He has worked with numerous school districts on curriculum and presented in-service workshops for middle and junior high school teachers on global geography, international understanding, and computer-assisted learning in geography. He has written world geography and regional geography texts for middle and junior high schools and was a design consultant for a global geography multimedia program for the middle schools. He had been active in the Commission on Geographic Education for the International Geographical Union. He has carried out field excursions and study in the United States and in forty different countries.

George Vuicich is a professor of geography at Western Michigan University and has taught geography since 1950 from grade six through the graduate level in Iowa, Wisconsin, Colorado, and Michigan. His B.A. (geography) and Ph.D. (geography) are from the University of Iowa. He was associate director of the High School Geography Project from 1965 to 1968. His teaching and research specializations include geographic education, urban geography, and quantitative methods in geography. He has conducted numerous workshops and institutes on teaching geography.

PREFACE TO THE REVISED EDITION

Before publication of the *Guidelines for Geographic Education: Elementary and Secondary Schools*, professional geographers met with little success in reforming and improving geographic education in the schools. The lukewarm reception of the High School Geography Project in 1965 and the decade following contrasted strongly with the warm educational and public reception to the *Guidelines* in 1984.[1] The *Guidelines* initiated a decade of progress and reform in geographic education.

A new era of educational reform may emerge with the imminent publication of voluntary National Standards Projects in geography, history, economics, and civics. NCSS (National Council for the Social Studies) has released curriculum guidelines for the social studies. These projects should profoundly influence the ways teachers teach and students learn these subjects. In the first edition of this book, I noted that if geography were strengthened in the schools, social studies education would benefit accordingly. If educators are serious about these new curricular initiatives and view them as ways to improve our schools and society, our future citizens will participate widely, rigorously, and intelligently in civic, social, and environmental decision making.

More than 15,000 copies of the first edition of *Strengthening Geography in the Social Studies* have been sold. Readers will find the material in that edition vital and surprisingly prescient. The principles, concepts, and ideas are still valid.

Curricula in geography and social studies have undergone significant changes since 1988. Geography and social studies educators have responded to the imperative of international educational competitiveness noted in the National Education Goals and have specified world class standards in geography and social studies. The following will chronicle and comment on the new geography standards and their relationships to geographic education, the *Guidelines for Geographic Education,* and the social studies.

The nation's governors and the Bush Administration in 1989 devised the National Education Goals at an education summit in Charlottesville, Virginia. These goals identified geography as one of the core subjects for schools in the United States. The National Education Goals affirmed the public awareness of and the urgency for a comprehensive geographic knowledge to understand the world. To support these goals and concerns, the geography community has produced *Geography for Life: National Geography Standards.*

Antecedents to the National Geography Standards Project

Scheduled for release in October 1994 (and for wide distribution by the National Geographic Society), *Geography for Life: National Geography Standards* (1994)[2] drew inspiration and guidance from the *Guidelines for Geographic Education: Elementary and Secondary Schools* (Joint Committee on Geographic Education 1984: 3-8, 22-23), which became a framework for *organizing concepts* to teach and learn geography in the schools during the past decade. The centerpiece of the *Guidelines* included five logically developed and complementary fundamental themes—**location:** position on the earth's surface; **place:** physical and human characteristics; **relationships within places:** humans and environments; **movement:** humans interacting on Earth or spatial interaction; and

regions: how they form and change. These themes combined with *essential geographic skills* became central for understanding geography and applying its concepts (see Chapter 1 in this volume).

The Joint Committee on Geographic Education of the Association of American Geographers (AAG) and the National Council for Geographic Education (NCGE) outlined the structure and purposes of a Geographic Education National Implementation Project (GENIP) to coordinate initiatives in geographic education in the United States. These organizations endorsed the *Guidelines* and encouraged their adoption in the schools, promoted the development of standards, and generated projects and materials to advance geography teaching and learning. Since 1986 the National Geographic Society's Alliance Network committed enormous resources for thousands of teachers from every state of the United States and parts of Canada to improve their knowledge of geography and update its teaching. Almost a decade later, the *Geography Assessment Framework for the 1994 National Assessment of Educational Progress (NAEP)* (National Assessment Governing Board 1993, 17-23) proposed three learning outcomes (*space and place, environment and society,* and *spatial dynamics and connections*) that would guide the assessment of geographic learning in grades 4, 8, and 12 in 1994.

Developing the National Geography Standards

The National Council for Geographic Education (NCGE) administered the National Geography Standards Project in coordination with the American Geographical Society (AGS), the Association of American Geographers (AAG), and the National Geographic Society (NGS).[3] These major American geography organizations developed the standards according to a broad-based consensus process involving thousands of teachers, scholars, international geography specialists, and citizens. The geography standards group also communicated regularly with the history, civics, and science standards projects as well as other disciplines related to geographic learning, e.g., economics, language and literature, social studies, and mathematics.[4]

The standards core writing group and teacher-writing committee also observed the United States Department of Education's criteria for national content standards projects, the National Council on Standards and Testing's recommendations, and the potential review criteria of the Technical Planning Group on the Review of Education Standards for the National Education Goals Panel (Technical Planning Group 1993).[5]

The Geographical Point of View

The geographical point of view lends itself to inter- and multi-disciplinary learning because it organizes knowledge according to specific perspectives, skills, and methods of study rather than examining and interpreting discrete phenomena, e.g., Chapter 6 on Chernobyl in this volume.[6]

Four major premises underlie the significance of geographic understanding for everyone. They are: *existential*—it is there, and people want to understand the intrinsic spatial nature of their surroundings on Earth; *ethical*—by emphasizing and explaining the interdependency of living things with the physical environment, geography can foster a respect for the essential and natural life-support systems on Earth. Geography, therefore, can help establish criteria for assessing new technologies and implications of their use for the environment and culture; *intellectual*—geography captures the imagination by focusing on exploration and adventure. Geography stimulates curiosity about the world and its inhabitants, places ideas in context, provides the essential perspectives, information, concepts, and skills to address local, regional, and global questions, and can help to overcome parochialism and ethnocentrism; and *practical*—at a personal level, geography improves decision making about choices for living, schools, public safety, and other social concerns. At the public, foreign, and international policy levels, it contributes vital knowledge about the accelerating interconnectedness of the world.

The Vision of the Standards

Geography, in brief, is *the study of people, places, and environments from a spatial perspective.* Geographically informed persons understand and appreciate the interdependent worlds in which they live from a local to a global scale. Further, the study of geography has practical value through the application of a spatial view to life situations. The vision in *Geography for Life: National Geography Standards* means enriching people's lives by bringing the vitality and meaning of places and spatial patterns into the nation's classrooms.

The geography standards use the term *spatial* to signify the specific viewpoint geography contributes to knowledge. Spatial refers to the tangible and intangible characteristics of *spaces on Earth's surface* (e.g., distance, direction, physical and cultural patterns and arrangements) that distinguish them from other *spaces* on Earth's surface and give them geographical relevance.

The National Geography Standards attempt to demonstrate how students, teachers, and society can achieve geographic competencies by becoming geographically informed. Teachers and learners using the standards can advance beyond simple place-name identification and interesting facts about people, places and their characteristics while taking advantage of the skills, perspectives, and attitudes of geography to examine *the spatial relationships among people, places, and environment.*

Overview of *Geography for Life: National Geography Standards*

Geography for Life outlines the critical components of geography, those integral characteristics of content, skills, and perspectives without which a distinctive geographic point of view is impossible. It specifies Geography Content Standards derived from the "Essential Elements of Geography" (the core framework of the standards) and embodies all of the foregoing information, provides rationales for each, and recommends what *a student should know and be able to do as a result of studying geography by the end of grades 4, 8, and 12.* Finally, the National Geography Standards describe the student *behaviors* teachers should expect if teachers and their students are to achieve geography standards at the end of grades 4, 8, and 12. Extensive supporting material for the document appears in a glossary, references and sources, and an index.

The Standards

Geography Content Standards, embodied in six Essential Elements, distill the indispensable knowledge a geographically informed person must understand and be able to use as a result of K-12 education. Each essential element encompasses two or more standards (18 in all) that further refine and amplify each element. The Essential Elements thereby integrate the content of geography with a set of geographic skills and essential perspectives from which people view the world (see also Boehm and Petersen 1994). The Essential Elements are: (1) Seeing the World in Spatial Terms; (2) Places and Regions; (3) Physical Systems; (4) Human Systems; (5) Environment and Society; and (6) Applying Geography.

Geographic Skills

Geographic skills are prerequisite tools and techniques for thinking geographically and are critical to attaining world class standards in geographic education. Similar to the rigorous steps in the scientific method, these skills are elaborated upon in Chapter 7 of this volume. Their nomenclature in *Geography for Life* differs slightly from the skills identified in Chapter 7.[7] These skills are: asking geographic questions; acquiring geographic information; organizing (and presenting) geographic information; analyzing geographic information; and answering geographic questions (developing and testing geographic generalizations).

These skills increase in sophistication and comprehensiveness during the course of each student's educational experiences. *Geography for Life* elaborates on how students can develop and apply these skills.

Geographic Perspectives

Geography for Life underscores the significance of perspectives in geographic education by elaborating on how people view the world from a variety of perspectives derived from their experience. People must learn to accept diverse ways of looking at the world. Perspectives also incorporate values, attitudes, and beliefs. When analyzing, evaluating, or trying to solve a problem, students must review a range of perspectives as well as have an awareness that perspectives can change.

Student Achievement Using Content Standards

Geography for Life discusses student achievement in interpreting and using geography and how teachers can interpret student achievement using the standards. According to the standards, students must demonstrate world-class learning behaviors in geography specific to the content standards, skills, and perspectives appropriate to their grade level.

Learning in geography increases in breadth, depth, and sophistication as students progress through school. The definitions of learning behaviors at grades 4, 8, and 12 illustrate how students manifest world-class competence using the content standards as benchmarks for assessment.

Internationally competitive standards in geography seek to motivate students and to set achievable levels for learning. Achieving these standards will challenge all students and teachers. Students and teachers will be able to apply and use these successful learning behaviors in every part of the curriculum.

The Challenge of the Standards

Geography standards offer the best opportunity for all students to learn essential information about Earth as our home. Effective geography instruction requires well-prepared teachers working with the best possible facilities and materials so that all students can achieve world-class standards. The National Geography Standards mean that teacher preparation programs must anticipate the knowledge, skills, attitudes, and strategies teachers will require for meeting the challenge of these rigorous standards.

Local school districts, college and university geography departments, and commercial textbook and materials developers all have specific and necessary roles in advancing the standards.

Parents can assist their children by using home, travel, recreation, sport, and cultural experiences to reinforce and expand their children's knowledge of geography. Helping students to make sense of their environment should be the goal of every geography classroom and of every parent. By gaining an intimate familiarity with the workings of their local environment through rich educational opportunities students will be able to explore and ask questions about places beyond their communities. This is the beginning of a successful geography program. Using the ideas in *Strengthening Geography in the Social Studies* can assist in achieving the standards outlined in *Geography for Life: National Geography Standards.*

Salvatore J. Natoli

REFERENCES

Boehm, Richard G. and James F. Petersen. "An Elaboration on the Fundamental Themes of Geography." *Social Education* 58 (April/May 1994): 211-218.

Gould, Peter. *Fire in the Rain: The Democratic Consequences of Chernobyl.* Baltimore, Md.: Johns Hopkins University Press, 1990.

Joint Committee on Geographic Education, AAG and NCGE. *Guidelines for Geographic Education: Elementary and Secondary Schools.* Macomb, Ill. and Washington, D.C.: National Council for Geographic Education and Association of American Geographers, 1984.

Geography Assessment Framework for the 1994 National Assessment of Educational Progress (NAEP). Washington, D.C.: National Assessment Governing Board, 1993.

National Geography Standards Project. *Geography for Life: National Geography Standards.* Washington, D. C.: forthcoming 1994.

Silberman, Charles E. *Crisis in the Classroom.* New York: Vintage Books, 1971.

Technical Planning Group for Goals 3 and 4 on the Review of Education Standards. Washington, D.C. : National Education Goals Panel, 1993.

*GENIP's organizational members are: American Geographical Society (AGS), Association of American Geographers (AAG), National Council for Geographic Education (NCGE), and the National Geographic Society (NGS). In 1987, GENIP published its first major publication, *K-6 Geography, Themes, Key Ideas, and Learning Opportunities* and a companion volume in 1989, *7-12 Geography, Themes, Key Ideas, and Learning Opportunities,* with the cooperation of Rand McNally. GENIP has also published a number of other publications for teachers and will soon issue, *Spaces and Places: A Manual for Geography Teachers.* The NCGE, AAG, AGS, and NGS have also produced a number of educational publications and projects for teaching and learning geography. The National Geographic Society's Geographic Alliance Network now includes all 50 states and parts of Canada. Geographers are now negotiating with the College Board to develop an advanced placement test for geography.

[1] Similar to that received by most of the national curriculum projects of the same period. See Silberman, Charles E. *Crisis in the Classroom.* New York: Vintage Books, 1971: 171.

[2] *Geography for Life: National Geography Standards 1994* can be obtained by writing to: Geography Education Standards Project, 1145 17th Street, NW, Suite 2500, Washington, DC 20036-4688.

[3] The United States Department of Education, the National Endowment for the Humanities, and the National Geographic Society provided funding for the geography standards writing project.

[4] The project writers also examined national and international teaching and learning materials, research on geographic learning, curriculum guidelines, pedagogical practices, and conferred with educational and content experts. The *Guidelines for Geographic Education* (Joint Committee on Geographic Education

1984) and the *Geography Assessment Framework for the 1994 National Assessment of Educational Progress* (National Assessment Governing Board et al. 1993) served as seminal documents and base points in their deliberations.

5 World class quality; focus on the most important knowledge and skills in geography; useful to the needs of schools and society; reflect broad consensus building with a widely inclusive review process; balance a set of enduring dimensions or polarities; reflect the current state of scholarship in the discipline; are clear to students, teachers, and parents; are specific enough to measure achievement; define the common core of what students must know and be able to do but are flexible enough to accommodate student, local, state, regional, and cultural diversity; and are suitable to and within the capabilities of students to learn at various age and grade levels.

6 The nuclear disaster at Chernobyl was not, in itself, a geographical phenomenon, but the subsequent atmospheric and terrestrial contamination had not only local but widespread, if not global, spatial consequences that continue to affect the health and quality of the environment in its immediate vicinity. In addition, it has changed attitudes toward peace-time use of nuclear power generating plants (See Peter Gould. *Fire in the Rain: The Democratic Consequences of Chernobyl*, Baltimore: Johns Hopkins University Press, 1990).

7 Changes are noted in parentheses.

CHAPTER 1

MODERN GEOGRAPHY

Salvatore J. Natoli and Charles F. Gritzner

We shall not cease from exploration, and the end of all our exploring will be to arrive where we started and to know the place for the first time.—T. S. Eliot.

Modern geography is a product of a number of traditions and practices. The term 'geography', coined by the Greek mathematician and cosmographer Eratosthenes (ca. 200 B.C.), means "writing about or describing the earth." Modern geographers also write about and describe the earth. One might question why we need to describe the earth since all parts of it have been discovered. In geography, however, discovery occurs every day because the earth is a dynamic entity that changes constantly and its people and landscapes constantly interact in new and different ways.

The goal of modern geographic education is to teach knowledge about the earth, to use that knowledge for personal enlightenment and development, and to apply it in making important personal decisions and in participating intelligently in societal decision making that affects our lives.

In the 18th and 19th centuries, schools in the United States showed a profound interest in teaching geography although textbooks used during this period contained maps and statements to describe and explain geographical ideas and regions of the world that were in many ways inaccurate. Yet geography was a staple part of the curriculum as the North American continent was explored and settled and as several national identities developed within its boundaries (see chapter 3).

Unfortunately, the last 50 years have been characterized by a decline in the teaching of geography as a separate school subject. Although many colleges and universities still have strong programs, a number of important colleges and universities have phased out their geography departments. The reasons for this decline are unclear, yet one might surmise that the unique physical and cultural endowments of the United States produced for us a geography—one of beneficence and bounty—that our people either took for granted or ignored (Natoli 1986).

Geography (or at least some semblance of it—ideas, concepts, and skills) still appears throughout the curriculum although it generally lacks identity as a separate subject until the late middle, or junior high, school years. When you ask most school persons why they do not teach geography, they will reply, "But we do! We have maps and globes in every classroom; we teach map and globe skills, or we teach it along with history, science, or current events. Or, we recommend it for our noncollege-bound students."

Cirrincione and Farrell, however, report on the favorable climate for introducing and expanding geography in the social studies curriculum, while observing that many social studies teachers continue to lack academic preparation in geography and have erroneous perceptions of the discipline (chapter 2).

GEOGRAPHICAL LITERACY

The appallingly low levels of geographical literacy in the United States indicate that geography instruction in the schools must be improved. Unfortunately, most of the tests for geographical literacy examine only place-name knowledge. Although place names are important in geography and no geographical knowledge can be obtained without knowledge or sense of place, simply knowing where places are is not geographical literacy. Ability to locate words in a dictionary does not make a person literate. Literacy begins when one is able to understand the meanings of words, employ them in coherent sentences and paragraphs, *and* understand clearly the meanings of these word combinations when reading them.

An atlas is sometimes misperceived as a dictionary of geography. It is rather an alphabet of geography, and only deep knowledge of geography can make it into a basic dictionary. Thus geographical literacy is attained only when people understand why places are where they are, what these places are like, and how they relate to these people and to other places. What appear to be measures of geographical literacy are indeed complex.

For example, when the world tried to understand the meaning of the Chernobyl nuclear disaster, it was important not only to know the location of the Chernobyl reactor (and very few persons did know) but to know why its location produced the particular pattern of fallout from the atmosphere it did during the spring of the year. The pattern surprised some people because Chernobyl lies in the path of the prevailing westerlies. Why should areas in eastern Europe be affected? Thus a basic knowledge of overall patterns in geography may be insufficient to explain local atmospheric patterns at different times of the year. In chapter 6, Morrill, Sellers, and Justham attempt to demonstrate the complexities of the patterns of nuclear fallout from the Chernobyl nuclear reactor and its short-term and long-term effects.

Just as literacy in any language requires a common core of knowledge derived from a common cultural heritage—both endogenous and exogenous—as well as an understanding of it and ability to communicate it effectively, so geographical literacy depends upon and contributes to the improvement of any population's cultural literacy. Of the more than 4,000 items listed in the supplement to E. D. Hirsch, Jr.'s *Cultural Literacy* (1987), almost 700 (more than 16 percent) are geographical. Thus one's cultural literacy is improved by knowledge of the geographical items as well as by improved understanding of and appreciation for the other historical and literary entries in the list.

Critics have accused Hirsch of ethnocentrism, but it is incontrovertible that one cannot achieve basic cultural literacy unless one first knows one's own culture. Americans will continue to have difficulty understanding the rest of the world unless they sharpen their powers of observation and knowledge of their own geography.

Like Alice, we will live in a constant state of wonder if we lack basic information and knowledge about how our own region (the United States) is different from or similar to other regions, why the seasons and climatic patterns change, or why earthquakes occur with greater frequency in some regions of the world than others.

MODERN GEOGRAPHY

Modern geography, unlike most other scientific and humanistic disciplines, is defined not so much by the objects it studies—nations, places, regions, the world—as by the methods and tools it uses to examine them. One of the realities of existence is that all "physical and social processes require space to operate" (Morrill 1985). Thus, it is the *spaces* in which events occur that are the focus of geographical study. Some events that occur spatially, like some events that occur temporally (in history), have greater significance than others and must therefore be subjected to careful and systematic scrutiny.

In the distant past, the major tools of the geographer were the compass, the sextant, and voyages of exploration. From these, the most important geographical tool, the map, was born. These early maps of discovery paved the way for increasingly accurate and orderly descriptions of the earth's surface. Today, the map is still the primary tool of geography, and today's maps demonstrate degrees of detail and sophistication unforeseen by early mapmakers. The range of technologies used for cartographic representation has improved immeasurably. Remote sensors and advanced computers have largely been responsible.

Yet modern geography still addresses similar questions to those it addressed in the past: Where, why, and of what significance are places on the earth's surface? The five themes of geography outlined in *Guidelines for Geographic Education: Elementary and Secondary Schools* (Committee on Geographic Education 1984) address these questions by providing a blueprint of the scope of geographical study. The interrelated themes—location, place, relationships within places (human-environmental relationships), movement (relationships between places or spatial interaction), and regions (how they form and change)—provide the essential elements for geographical study (see insert of Fundamental Themes in Geography).

Fundamental Themes in Geography

LOCATION:
Position on the Earth's Surface

Absolute and relative location are two ways of describing the positions of people and places on the earth's surface.

PLACE:
Physical and Human Characteristics

All places on the earth have distinctive tangible and intangible characteristics that give them meaning and character and distinguish them from other places. Geographers generally describe places by their physical or human characteristics.

RELATIONSHIPS WITHIN PLACES:
Humans and Environments

All places on the earth have advantages and disadvantages for human settlement. High population densities have developed on flood plains, for example, where people could take advantage of fertile soils, water resources, and opportunities for river transportation. By comparison, population densities are usually low in deserts. Yet flood plains are periodically subjected to severe damage, and some desert areas, such as Israel, have been modified to support large population concentrations.

MOVEMENT:
Humans Interacting on the Earth

Human beings occupy places unevenly across the face of the earth. Some live on farms or in the country; others live in towns, villages, or cities. Yet these people interact with each other: that is, they travel from one place to another, they communicate with each other, or they rely upon products, information, and ideas that come from beyond their immediate environment.

The most visible evidences of global interdependence and the interaction of places are the transportation and communication lines that link every part of the world. These demonstrate that most people interact with other places almost every day of their lives. This may involve nothing more than a Georgian eating apples grown in the state of Washington and shipped to Atlanta by rail or truck. On a larger scale, international trade demonstrates that no country is self-sufficient.

REGIONS:
How They Form and Change

The basic unit of geographic study is the region, an area that displays unity in terms of selected criteria.

We are all familiar with regions showing the extent of political power such as nations, provinces, countries, or cities, yet there are almost countless ways to define meaningful regions depending on the problems being considered. Some regions are defined by one characteristic such as a governmental unit, a language group, or a landform type, and others by the interplay of many complex features. For example, Indiana as a state is a governmental region, Latin America as an area where Spanish and Portuguese are major languages can be a linguistic region, and the Rocky Mountains as a mountain range is a landform region. A geographer may delineate a neighborhood in Minneapolis by correlating the income and educational levels of residents with the assessed valuation or property and tax rate, or distinguish others by prominent boundaries such as a freeway, park, or business district. On another scale we may identify the complex of ethnic, religious, linguistic, and environmental features that delineate the Arab World from the Middle East or North Africa.

These themes can be explained at very basic levels of the elementary schools and then handled at higher and higher levels of sophistication as students progress to higher grades. Of these themes, the region expresses best how geographical knowledge can be synthesized. According to John Fraser Hart (1982), the region is the "highest form of the geographer's art." Some geographers during the past three decades have relegated regional geography to a very low priority in their research. Others have bemoaned its demise and felt that geographers have lost their birthright as a result. By neglecting the region, geographers have neglected a central focus of their traditional scholarship. Hart (1982), however, argued for the return to regional studies in geography to stem the tide in the growth of systematic subfields in favor of a unifying theme for the discipline that ties together all the disparate phenomena with which geographers deal.

Contrary to appearances, regional geography has not disappeared, though it has been altered. In the early part of the 20th century, the regional geographer's stock-in-trade was to describe the unique characteristics of areas (regions). The reaction to this type of descriptive geography triggered a revolution in the 1950s and 1960s that attempted to redefine geography as a spatial science, directed toward the search for general laws of spatial structure (Garrison 1979; Gould 1979).

> The uniqueness of regions is now to be understood as socially constructed, the result of a synthesis of place-specific characteristics with more general social processes; reciprocally, how these more general processes unfold is influenced by regional specificities. (Hudson 1987)

Modern geography has emerged not only by the rigorous application of the canons necessary for scientific revolution but out of the crisis conditions of the Depression Era. During the 1930s, geographers were recognized as useful—as having the capabilities to assist in solving such problems of the environment as soil erosion and watershed management, a host of human and natural resource problems, and the social and physical ramifications of these problems.

APPLIED GEOGRAPHY

During World War II, the usefulness of geographers grew by virtue of their abilities in cartographic representation, in map and aerial photo interpretation, and in knowledge of foreign areas. Many geographers served the war effort in intelligence agencies and in strategic and logistical planning (Natoli 1986).

In the postwar years, the growth of opportunities in urban, regional, and rural planning provided still more outlets for geographical skills. The increasing use of geographers in business, environmental management, and in problems of economic development was further expression of the utility of geographical methodologies. Geographers also participated in studies that led to the rejuvenation of war-torn Europe and Asia. Today, geographic research and applications provide essential studies for planning economic and social development in the Third World.

GEOGRAPHY AS A SEPARATE SUBJECT

The specific utility of the geographical approach to real-world problem solving argues for teaching geography as a separate subject. Although other social sciences, physical sciences, and the humanities ostensibly deal with geographical content, they tend to ignore important spatial aspects of issues and problems. By emphasizing the geographical point of view in addressing these problems, separate and distinctive geography courses reduce the chances of ignoring or mistreating these important concepts (chapter 2).

Another argument for geography's separate identity in the curriculum is that few of the traditional disciplines offer as many real-world situations for students to study that may eventually provide them with skills for potential careers. Aside from the extremely favorable job market for cartographers—especially for those trained in computer cartography and for individuals with special competencies in interpreting and analyzing remotely sensed imagery—training in geography offers a wide variety of marketable skills. These include area analysis for planning social and resource use activities that provide information and intelligence on foreign areas; projecting the environmental impacts of public and private developments on both natural and human environments; location analysis for selecting optimal sites for public and private service installations; research, evaluation, and presentation that can be applied directly to local, state, and federal government problems; and, most important, the ability to link diverse fields of knowledge and understand information from a wide variety of sources (Department of Geography, University of Illinois 1982).

CAREERS IN GEOGRAPHY

A career in geography might appeal most to those who wish to examine social and environmental issues and problems from the broad perspective of geography or even from a range of various perspectives for seeking alternative solutions. The combination of natural and social science skills and perspectives permits the geographer to view and analyze complex issues and problems. For example, one of today's most pressing problems in the United States as well as in other industrialized countries of the world is to seek safe sites for disposing nuclear waste products. Geographic Information System (GIS) specialists are performing some of the most intricate studies of complex human and environmental relationships. The analysts combine and reformulate data from literally hundreds of human and physical variables in order to select several alternative "optimal" sites to dispose of toxic nuclear wastes. A subproblem is to determine the most environmentally safe transportation routes to carry these products from their sources to the disposal site. The sites and routes finally selected for the disposal and transportation of the wastes will depend upon the affected communities' willingness to permit either the location of these wastes in, or their passage through, their vicinities. These and less dramatic environmental problems occupy the time of many professional geographers. Academic geographers also contribute theories and applications for geographical problem solving.

GEOGRAPHY INTEGRATED WITHIN THE
SOCIAL STUDIES

Stoltman and Libbee examine the development of geography as part of the social studies in chapters 3 and 4. They offer several models for geography's effective incorporation into the curriculum. These models provide an acceptable alternative to geography as a separate subject and demonstrate how important it is that social studies teachers recognize the power and significance of geographic ideas and concepts when they occur in the curriculum. In addition, these models evidence meaningful modes of articulation within the curriculum.

Morrill, Sellers, and Justham (chapter 6) and Vuicich, Stoltman, and Boehm (chapter 7) develop two somewhat related but contrasting approaches for incorporating or stengthening geography in the social studies curriculum. Morrill et al. demonstrate a mode of analysis for studying the geographical signficance of the Chernobyl nuclear disaster using the five fundamental conceptual themes of geography (Committee on Geographic Education 1984). Vuicich et al. emphasize how students' use of geographic skills is essential for social problem solving and for enhancing learning in the social studies.

HOW TO BEGIN

If the important issue in this volume is to strengthen geography in the social studies curriculum, then we must examine some fundamental steps schools must take before this can happen. Strengthening geography cannot occur by administrative fiat, although school administrators must provide both moral and financial support. Eventually, it must be the classroom teacher who will implement the program.

Teachers become anxious when asked to develop new programs in their schools, especially if they lack sufficient academic background in the discipline to be developed. A recent survey conducted by Cirrincione and Farrell (chapter 9) provided clear evidence that social studies teachers in the United States look favorably upon increasing the content and quality of geography in their schools. A favorable atmosphere is not enough unless both in-service and preservice teacher preparation in geography improves. Poorly prepared teachers will undermine any long-range goal to improve geographic education in the schools.

In chapter 5, Spetz summarizes the work of a National Council for Geographic Education (NCGE) and Association of American Geographers (AAG) committee that developed guidelines for teacher preparation in geography. Although the guidelines say little about in-service training, we can assume what background might be necessary for minimal competency to teach geography.

In chapter 8, Marran, Natoli, and Juliette offer both theoretical and practical advice for beginning or improving a geography program in elementary and secondary schools. In addition, they detail space and equipment needs for supporting geography programs and in turn strengthening the social science and natural science programs.

TOWARD GEOGRAPHICAL LITERACY

The purpose of this volume then is to assist teachers in eradicating geographical illiteracy in our society. Earlier in this chapter, we discussed the importance of geographical literacy as it related to cultural literacy. A few years ago, Gritzner (1981) mused about the sorry state of geographical illiterates:

> To individuals lacking a global "mental map" the world must be little more than a confusing hodgepodge; places without location, quality, or context; faceless people and cultures void of detail, character, or meaning; vague physical features and environments for which terminology, mental images, causative agents and processes, and human patterns are lacking; temporal events that occur in a spatial vacuum; and a host of critical global problems for which they have no criteria on which to base analyses, judgments, or attempts at resolution. Such individuals are prisoners of their own ignorance or provincialism.

Geographical literacy demands that all students gain a common knowledge of their immediate and world environments. One important characteristic of geography is its concern for the earth as an ecological system. This equips us with a global perspective for analyzing world problems. By studying regions of the changing world, students begin a lifelong learning process with a rich human and physical context.

Geographical literacy attempts to explicate the sentences of geography and distinguish them from the words (place names) that the public equates with geographical knowledge. Such knowledge cannot substitute for the coherence of thought and meaning in a geographically complete sentence. Geographical knowledge might diminish in our students a sense of wonder about the world in which they live. On the other hand, knowledge of geography should stimulate students' curiosities about the wonders as well as the problems of the world in which they live. It also might help to cultivate in them a sense of stewardship for the fragility of many of the earth's environments. Such knowledge about and appreciation of the world can lead students to satisfying lives and improve their participation as citizens in this democratic society and as partners in the world community.

We can no longer sanction the practice of isolating ourselves from the global community. As each year passes, our relationships in the complex mosaic of nations become an increasingly inextricable part of our passage through life. Knowledge of geography will improve the meaning, safety, and enjoyment of this journey.

REFERENCES

Committee on Geographic Education, National Council for Geographic Education (NCGE) and Association of American Geographers (AAG). *Guidelines for Geographic Education: Elementary and Secondary Schools.* Washington, D.C., and Macomb, Illinois: AAG and NCGE, 1984.

Department of Geography, University of Illinois. *Applied Geographers Can Help Increase Efficiency and Effectiveness of Your Agency.* Washington, D.C.: Association of American Geographers, 1982.

Garrison, William. "Playing with Ideas." *Annals of the Association of American Geographers* 69 (March 1979): 118–20.

Gould, Peter. "Geography 1957–1977: The Augean Period." *Annals of the Association of American Geographers* 69 (March 1979): 139–50.

Gritzner, Charles F. "Geographic Education—Where Have We Failed?" *Journal of Geography* 80 (December 1981): 264.

Hart, John Fraser. "The Highest Form of the Geographer's Art." *Annals of the Association of American Geographers* 72 (March 1982): 1–29.

Hirsch, E. D., Jr. *Cultural Literacy: What Every American Needs to Know.* With an appendix, "What Literate Americans Know," by E. D. Hirsch, Jr., Joseph Kett, and James Trefil. Boston: Houghton Mifflin, 1987.

Hudson, Ray. "Flying the Flag for Flagging Geography." *The Times Higher Education Supplement,* October 23, 1987.

Morrill, Richard. "Some Important Geographic Questions." *The Professional Geographer* 37 (August 1985): 263–70.

Natoli, Salvatore J. "The Evolving Nature of Geography." In *Social Studies and Social Sciences: A Fifty-Year Perspective,* NCSS Bulletin No. 78, edited by Stanley P. Wronski and Donald H. Bragaw. Washington, D.C.: National Council for the Social Studies, 1986, 28–42.

CHAPTER 2

THE STATUS OF GEOGRAPHY IN MIDDLE/JUNIOR AND SENIOR HIGH SCHOOLS

Joseph M. Cirrincione and Richard T. Farrell

The need to improve our geographic knowledge and skills has increased steadily during the past several decades. There is substantial evidence that students have a limited understanding of basic geographic content. Despite this mounting evidence, curriculum developers have been less than enthusiastic about formally integrating geography into the curriculum. The role of geography from middle schools to high schools remains poorly defined and of limited educational value. Numerous professional and popular publications make these points clearly. In a recent NCSS Bulletin, for example, Winston (1986) pointed out that, "(1) U.S. children and many adults are deficient in geographic knowledge, skills and affective learning; and (2) geography is less significant than it might be as a component in social studies." Natoli (1986) was more blunt: "Many Americans are abysmally ignorant not only of the rest of the world's geography but of their own as well."

Such conclusions by professionals and public critics have not gone unnoticed. Numerous activities are under way across the United States to improve geographic education in the schools (see Introduction).

Throughout the United States, state geographical associations and committees have been organized or rejuvenated. In a few states, geography is now a requirement for graduation from high schools, and at least one state has identified geographic competence as one of six major goals of the social studies curriculum. Within the public sector, a Southern Governors Conference report recommended, among other items, that "Geography be taught as a distinct subject in kindergarten through the twelfth grade rather than the common practice of teaching it as part of a social studies course, if at all" (Vobejda 1986).

The Study

This is a preliminary study on some of the curricular issues facing professional geographers and educators. It focuses on experienced classroom social studies teachers' attitudes toward geography education. It attempts specifically to address four basic questions that educators should consider before initiating significant curriculum and educational reforms. First, are social studies teachers prepared adequately to assume the additional academic responsibilities for increasing the role of geography in the curriculum? Second, how do teachers feel about the current status of geography in the curriculum, and do they agree with the education critics that reform is necessary? Third, do teachers agree generally on how geography can be integrated effectively into the curriculum? Fourth, to what extent do various social studies teachers agree on major objectives of a geography program? Answers to these questions may appear self-evident, but there are almost no data to support such an assumption. There is, we believe, a critical need to develop a continuing data base on teacher preparation in geography and on the significant components in its content.

The Instrument

The authors designed a questionnaire and sent it for review to professional geographers and social studies educators. The revised questionnaire was then mailed to 1,138 social studies teachers. NCSS provided names and addresses from a national stratified random sample based on state ZIP codes. Responses were obtained from 594 teachers (52.1 percent). Of the 594, eighteen were rejected for various reasons. Usable responses totaled 576.

The instrument included three general parts. Part I dealt with teaching responsibilities and the nature and extent of the academic preparation of respondents in geography. Part II attempted to assess teachers' views on the role of geography in the social studies curriculum. Part III asked respondents to evaluate some broad goals of geography programs and some specific content items obtained from a review of six recent geography textbooks.

Profile of Teachers

The majority of the respondents (351, or 61 percent) were senior high school teachers. Seventy-five (13 percent) taught in junior high schools; sixty-four (11 percent) in middle schools. Other respondents identified themselves as supervisors (76, or 13 percent) or college instructors (7, or 1 percent). Three respondents failed to indicate their positions. A majority (51 percent) considered history as their primary teaching field. Thirteen percent were civics teachers; 10 percent indicated that geography was their primary teaching responsibility. The remainder taught a combination of social studies courses and were not classified by a specific teaching field.

Academic Preparation of Teachers

A potential problem in expanding the role of geography in the curriculum is that social studies teachers are probably not prepared adequately to assume the additional responsibility of teaching geography. Figure 1 presents results of this survey on the formal academic preparation of teachers in geography. To simplify the presentation of data, three categories were used: "0" no formal course; "1–3" formal courses; and ">3" more than three courses.

Figure 1
PROFILE OF TEACHER TRAINING IN GEOGRAPHY (N = 576)

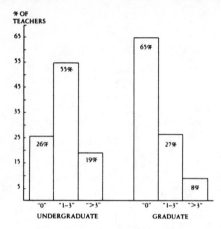

Table 1 suggests that social studies teachers generally have had limited academic preparation in geography. At least one-fourth had no undergraduate academic coursework in geography. Although the "1–3" column may indicate otherwise, 40 percent of this group had only one course. Of the 576 respondents, 48 percent (276) have had no more than one undergraduate course in geography. At the graduate level, 64 percent have had no advanced training. Although a substantial majority of the respondents were not geography teachers, but because all of them were social studies teachers, one might anticipate greater exposure to the field than was the case. The survey results indicate that history teachers have minimal preparation in geography despite geography's importance to understanding much of history's content. At the undergraduate and graduate levels, there is clearly a need to expand the amount of preparation social studies teachers receive in geography regardless of its role or organization in the curriculum.

In addition, the survey attempted to determine the academic background of respondents in geography and its subfields. The results are summarized in Figure 2. Respondents were asked to indicate their formal undergraduate or graduate level course work in any of the following five subfields of geography: physical, cultural, economic, world, or regional. As expected, the largest percentage of respondents had taken at least one course in physical geography. A partial explanation for the popularity of this course is that some states require physical geography for social studies certification; some universities permit it to satisfy a physical science requirement in the liberal arts or general studies programs. More likely, however, in colleges and universities that offer only one course in geography, they tend to define it almost exclusively as physical geography.

More surprising were the high percentages of social studies teachers who had no formal training in the geography subfields in which one would assume they would have some interest. Table 1 confirms this conclusion. History

Figure 2
TRAINING IN FIELDS OF GEOGRAPHY

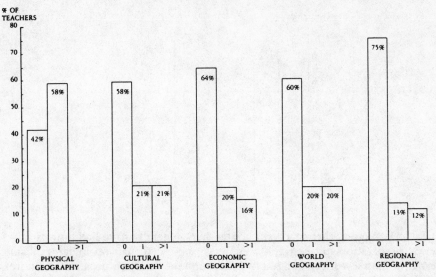

teachers, for example, who might be expected to have a greater interest in cultural or world/regional geography courses than in physical geography courses or civics teachers who might be expected to have a greater interest in economic geography courses clearly favored the physical geography course. Geography teachers also overwhelmingly favored physical geography (75 percent having had at least one course). In fact, physical geography was the only subfield in which teachers in all the teaching areas had taken a course. The popularity of

Table 1.
PERCENTAGE OF TEACHERS BY SUBJECT FIELDS HAVING TRAINING IN DIFFERENT AREAS OF GEOGRAPHY

	PHYSICAL GEOGRAPHY			CULTURAL GEOGRAPHY			ECONOMIC GEOGRAPHY			WORLD/ REGIONAL GEOGRAPHY		
	0	1	>1	0	1	>1	0	1	>1	0	1	>1
HISTORY TEACHERS	47%	53%	-	63%	21%	16%	68%	19%	13%	68%	17%	15%
GEOGRAPHY TEACHERS	25%	75%	-	38%	26%	39%	49%	20%	31%	54%	19%	27%
CIVICS TEACHERS	34%	60%	-	65%	22%	13%	60%	26%	14%	67%	22%	11%
OTHERS	44%	56%	-	53%	17%	30%	65%	15%	20%	80%	6%	14%

physical geography is undoubtedly the result of certification requirements. Because many states do not specify geographic content to satisfy certification requirements and because physical geography is one of the most frequently taught introductory courses, yet no teachers took a second course, perhaps because physical geography becomes increasingly specialized after the introductory course.

Another unusual finding was the apparent lack of teacher preparation in world/regional geography courses, especially considering the current emphasis on international/multicultural education and the traditional emphasis on world/regional geography in the school curriculum. An unusually high percentage of the respondents (75 percent) had no preparation in regional geography. The teaching field of the respondents did not correlate with their geography preparation and only slight variations were evident among the four groups. Predictably, more geography teachers than other teachers surveyed had more than one world/regional course.

This apparent lack of interest in world/regional geography could reflect the nature of college and university geography programs. The respondents averaged 18 years of teaching. One could assume, therefore, that they received most of their collegiate preparation when regional geography was declining as an emphasis at colleges and universities with a corresponding increase in topical/systematic geography. Today programs continue to follow introductory physical, economic, and cultural geography courses with advanced work in topical/systematic geography courses. Although there is some renewed interest in regional courses, most colleges and universities list them as single course electives.

Teachers' Attitudes toward the Status of Geography

Part II of the survey instrument attempted to assess teachers' attitudes toward the role and status of geography in the social studies curriculum. Only 33 percent of the respondents indicated that some type of geographic instruction was required. This may be an inflated figure because some middle or junior high school teachers indicated that they considered world cultures courses as geography courses or that they taught geography independently but as a series of units within a required social studies course. Examination by teaching levels indicated that a much higher percentage of schools required geography at the middle/junior high school level (60 percent) than at the high school level (20 percent). This seems to confirm that geography plays a limited role in the high school curriculum and that teachers view it primarily as a middle/junior high school level course.

When asked whether the offerings in geography should change, all respondents indicated a high level of agreement for expanding the offerings.

Table 2 shows that more than 70 percent of all the respondents regardless of grade level or teaching field favored expanding geography offerings. Senior high school teachers supported expanding offerings more than any other group. Although supervisors did not favor reducing offerings in geography, they constituted the lowest percentage of those advocating expansion and the highest percentage for supporting the status quo. Thus, both education critics and

Table 2. TEACHERS' ATTITUDES TOWARD CURRENT OFFERINGS IN GEOGRAPHY

	EXPAND	SUFFICIENT	REDUCE
ALL GRADE LEVELS	75%	23%	1%
MIDDLE/ JUNIOR HIGH	74%	23%	-
SENIOR HIGH	76%	23%	1%
SUPERVISORS	71%	29%	-
(FIELDS)			
CIVICS	81%	18%	-
GEOGRAPHY	71%	29%	-
HISTORY	75%	22%	2%
OTHERS	74%	25%	-

teachers agree that geography offerings should be expanded in the curriculum. One respondent observed that geography should be introduced at the elementary levels and that "it needs to be a progressive system that continues through high school."

Teachers' Attitudes toward the Organization of Geographic Instruction

One of the persistent questions facing curriculum developers when attempting to expand the role of geography in the social studies curriculum is whether to integrate the subject into existing courses or offer it as a separate subject. Typically, educators at the middle/junior high school level view geography as a separate course. The survey tends to support this conclusion. At the middle/ junior high school levels, 53 percent of the respondents reported that geography was taught as a separate course. At the high school level, 47 percent reported the separate course option. Twenty-nine percent of all respondents reported that geography was taught as a series of units in other social studies courses. More middle/junior high teachers (35 percent) favored the latter option than high school teachers (26 percent). Fifteen percent of the middle/junior high

Table 3. TEACHERS' ATTITUDES ON THE ORGANIZATION OF GEO-
GRAPHIC INSTRUCTION

	SEPARATE COURSE	INTEGRATED INTO HISTORY	UNITS IN SOCIAL STUDIES
ALL GRADE LEVELS	54%	29%	17%
MIDDLE/ JUNIOR HIGH	49%	30%	21%
SENIOR HIGH	54%	30%	15%
SUPERVISORS	56%	24%	19%
(FIELDS)			
CIVICS	60%	16%	20%
GEOGRAPHY	83%	11%	6%
HISTORY	46%	40%	13%
OTHERS	53%	17%	28%

school and 18 percent of the senior high school teachers sampled reported that they were given the discretion to teach geography as a separate course or as units within other courses. Although a high percentage of the teachers supported the separate course organization, it is telling that a sizable portion of the respondents reported that it was simply treated as a series of units in the curriculum.

Respondents were then asked how they preferred to see geography courses organized. They were given three options: taught as separate courses, integrated into history courses, or taught as separate units in other social studies courses. The results are reported in Table 3.

Among respondents teaching at different grade levels, there were only slight differences in their attitudes toward the organization of geography in the curriculum. Clearly more than half favored separate courses; nearly a third supported integration into history courses. It is noteworthy that a high percentage of teachers at all levels continue to believe that geography should be taught as a separate unit in other courses.

An analysis of the same question by teaching fields presents a more diverse pattern. Unlike the attitude previously expressed, teachers' subject fields strongly

17

influence their attitudes. Geography teachers overwhelmingly opted for separate courses (83 percent), whereas less than half (46 percent) of the history teachers favored this organization. As expected, a rather high percentage (40 percent) of the history teachers favored integrating the two subjects. That more history teachers (46 percent) supported separate courses than integration (40 percent) suggests no clear direction for curriculum reformers. Civics and "other" respondents' rather high percentage (over 20 percent) in favor of separate units in other courses probably reflects existing arrangements in most schools.

Finally, to gauge the general attitude of social studies teachers toward geography, respondents were asked if they were willing to teach a course. As in the case of expanding geography's role in the curriculum, an overwhelming majority (73 percent) of all respondents indicated a willingness to teach geography. More middle/junior high school teachers (82 percent) favored the idea than high school teachers (70 percent).

Teachers' Attitudes toward Selected Objectives and Content in Geographic Instruction

The final section of the questionnaire attempted to identify general aspects of geography that social studies teachers perceived as important and to determine the level of agreement, if any, among teachers in different social studies fields. First, we selected seven general objectives associated with geographic instruction as representative (Maryland State Department of Education 1984). Teachers were asked to rate the objectives on a Likert scale ranging from (1) "not important" to (5) "very important." The higher the score for an objective, the greater the degree of importance teachers attributed to that aspect of geography.

The most obvious observation is that respondents in all subject fields tended to rank each objective nearly the same (Table 4). All four groups ranked location of major cultural and physical features first or second. Considering the traditional impression that historians are preoccupied with a narrow view of geography—location of places—they ranked "world interdependence" first and location second in this study. Geography teachers gave "world interdependence" only a fourth place ranking, and, more important, they rated the physical environment and economic activities higher. As might be expected, civics teachers gave slightly greater importance to "major economic factors" than the other groups. One of the most important conclusions the data seem to support is the low ratings given objectives four (population patterns), five (transportation and communication networks), and seven (respect for the environment). All groups apparently fail to recognize the relationships among world interdependence and population patterns and transportation and communication.

Using the same scale, teachers were asked to rate the level of importance of sixty specific items of content obtained from a content analysis of six current geography textbooks. Normal procedures for content analysis were used including reliability tests. Specific items were taken from each broad area of geography. Table 5 presents a summary of the responses. Specific items from the survey were grouped together according to the broad subfields of geography and a

Table 4. TEACHER RESPONSE TO SAMPLE GEOGRAPHY OBJECTIVES FOR THE SOCIAL STUDIES CURRICULUM.

1. NOT IMPORTANT
2. MARGINALLY IMPORTANT
3. IMPORTANT
4. GENERALLY IMPORTANT
5. VERY IMPORTANT

	ALL		(TEACHING FIELDS) CIVICS		GEOGRAPHY		HISTORY		OTHERS	
Objective	MEAN	RANK	MEAN	RANK	MEAN	RANK	MEAN	RANK	MEAN	RANK
1. Locate and interpret major cultural and physical features.	4.48	(1)	4.48	(1)	4.56	(1)	4.44	(2)	4.50	(2)
2. Understand the interaction of culture and technology in the use and alteration of the physical environment.	4.35	(3)	4.29	(4)	4.45	(2)	4.26	(3)	4.47	(3)
3. Understand the characteristics of major economic activities and the factors influencing their location.	4.29	(4)	4.30	(2)	4.36	(3)	4.22	(4)	4.33	(5)
4. Understand patterns of population growth and settlement in different cultures and environments.	4.03	(6)	4.02	(6)	4.01	(6)	3.96	(6)	4.12	(6)
5. Understand the role and impact of transportation and communication in linking people and environments.	3.96	(7)	3.90	(7)	4.00	(7)	3.89	(7)	4.07	(7)
6. Understand and appreciate various dimensions of world interdependence.	4.47	(2)	4.30	(2)	4.35	(4)	4.46	(1)	4.60	(1)
7. Demonstrate respect for the environment.	4.23	(5)	4.05	(5)	4.30	(5)	4.18	(5)	4.39	(4)

Table 5. SUMMARY OF TEACHERS' RESPONSES TO THE IMPORTANCE OF INDIVIDUAL ITEMS OF CONTENT

	MEAN	RANK
Physical	3.40	(7)
Skills	3.78	(3)
Cultural	4.07	(1)
Economic	3.69	(4)
Population	3.66	(5)
Movement	3.37	(8)
Urban	3.59	(6)
Global	3.81	(2)

summary mean score was generated for each area. Items included under the Global Dimension were selected without regard to the subfield to which they were originally assigned. The common characteristic was their emphasis on global interdependence.

Several observations seem in order. First, there was a high level of agreement in the way teachers rated the broad objectives with related items. Whether responding to the objectives or the items, the respondents were generally consistent and placed the greatest value on culture and world interdependence— the least on transportation and communication (movement). Second, they rated skills high, third in importance. This suggests that, although teachers think teaching geographic skills is important, it is not their sole or primary responsibility as some geography advocates would imply.

The next three categories tended to cluster, reflecting the common thread inherent in the content. All three generally place a heavy emphasis on economic rather than cultural topics. Physical geography was rated low. This is puzzling because it is the subfield in which most teachers had some preparation, and its content most clearly separates geography from other social studies courses. Movement received the lowest rating. Perhaps this results from a misunderstanding of the concept and how it relates to geography and the social studies curriculum.

Conclusions and Observations

Our data suggest the following:

1. If geography is to assume a significant role in the social studies curriculum, the inadequacy of formal preparation in geography of teachers must be addressed. At present, teachers lack both the breadth and depth of background to handle the necessary content.

2. Teachers seem to disagree on the best ways to organize geography programs for the most effective teaching and learning. The question of a separate geography course or its integration into other courses remains unsettled. This is an important question both in terms of concepts, skills, and subfields of geography to be stressed and their appropriate grade levels.

3. Teachers are not obstacles to expanding geography in curriculum. The respondents were willing to teach a course in geography and strongly supported increasing its role in the curriculum.

4. Although poorly prepared in cultural geography, social studies teachers recognize its importance and look to geography to deal with the cultural dimensions of the social studies curriculum.

5. Social studies teachers appear to endorse the view that geography teachers should assume the major responsibility for teaching global education and world interdependence.

6. Teachers need to learn about new directions in geography through alternative delivery systems, in-service programs, and postbaccalaureate course work.

7. Although critics of the quality of geographic education tend to give primary emphasis to skill development and generally define geography in terms of skills, teachers tend to recognize the importance of skills but give greater emphasis to conceptual knowledge in geography. Improving geographic skills alone is insufficient for improving the geography curriculum.

Teachers must assume clear and active roles in any effort to revise geography's role in the social studies curriculum—particularly in defining the goals and objectives of geographic instruction. This cannot occur unless all those interested in advancing the field—teachers, educators, and professional geographers—cooperate.

REFERENCES

Committee on Geographic Education. National Council for Geographic Education and Association of American Geographers. *Guidelines for Geographic Education: Elementary and Secondary Schools.* Washington, D.C., and Macomb, Ill.: AAG and NCGE, 1984.

Maryland State Department of Education. *Social Studies: A Maryland Framework.* 1984.

Natoli, Salvatore J. "The Evolving Nature of Geography." In *Social Studies and the Social Sciences: A Fifty-Year Perspective,* NCSS Bulletin No. 78, edited by Stanley P. Wronski and Donald H. Bragaw. Washington, D.C.: National Council for the Social Studies, 1986.

Vobejda, Barbara. "U.S. Students Called Internationally Illiterate." *The Washington Post,* November 22, 1986, Section A, p. 7.

Winston, Barbara J. "Teaching and Learning in Geography." In *Social Studies and the Social Sciences: A Fifty-Year Perspective,* NCSS Bulletin No. 78, edited by Stanley P. Wronski and Donald H. Bragaw. Washington, D.C.: National Council for the Social Studies, 1986.

CHAPTER 3

GEOGRAPHY WITHIN THE SOCIAL STUDIES CURRICULUM

Michael Libbee and Joseph Stoltman

Both geography and the social studies have been characterized by a fuzzy image and competing academic traditions. Recent work in social studies and geography has served to clarify their respective roles in the school system, as well as their disciplinary relationships. This chapter presents a brief history of geography in the schools and describes how different traditions of geography correspond with different traditions in the social studies.

Historical Perspective on Geography in the Curriculum

Geography has been part of American education since the 17th century when Harvard University introduced map and globe study (Warntz 1964). The elementary schools of the time also found geography ideally suited to the intellectual activities and abilities of children, often including stand-and-recite memory exercises. Elementary and grammar school teachers were devoted to memorization exercises that enabled their students to pass college admissions examinations in geography successfully (Warntz 1964). Colleges and universities often viewed geographic knowledge as an important requirement for entering students.

Admissions requirements to higher education in the early 1800s resulted in teaching increasingly greater quantities of geography in both elementary and secondary schools (Boyles 1926). After about 1830, the importance of geographic knowledge was recognized further through the enactment of state laws requiring the teaching of geography in the elementary schools and, in some instances, the study of the geography of the home state (Rumble 1946).

Early Textbooks in Geography

Early geography textbooks guided school geography to focus clearly upon the acquisition of facts. The earliest American text, *Geography Made Easy,* published by Jedidiah Morse in 1784, was written so that almost anyone could teach geography. It measured pupils' progress by their ability to memorize standard answers to standard sets of questions. The early editions of the Morse "geographies" contained numerous inaccuracies, had a strong bias in favor of New England, reflected a strong religious orthodoxy (Morse was a minister), and took an extremely conservative position on morality (Warntz 1964).

Geography in the Elementary Schools

Definitive accounts of geography and its role in shaping the elementary school curriculum in the United States are not available. The best sources are general treatments of the history of curriculum development that provide overviews of the disciplines and processes involved. Nearly all such general treatments stress the close relationship between school and society that was reflected in the elementary school curriculum from an early period (Goodlad and Share 1973). The most comprehensive research on the role of geography in the elementary school curriculum has been reported by James (1962) and Vuicich and Stoltman (1974). Geography in the elementary grades appears consistently as rote description and memorization of geographic detail. The elementary focus on facts also led to criticism, since it resulted in colleges' and universities' disparaging the intellectual legitimacy of geography (Warntz 1964).

Geography in the Secondary Schools

The period from 1820 to 1870 was especially important in American secondary education. It was especially important for geography since the Swiss educator, Johann Pestalozzi, was promoting the philosophy that direct observation and sense perception were essential to meaningful learning by children. Those concepts began to appear in the United States as attractive and interesting books for students. Maps and line drawings became essential parts of texts. American educators proposed that map drawing, the study of the home region through direct observation, and experimentation using globes were the most effective ways to teach geography (Rosen 1957). The appearance of geography in the secondary curriculum coincided fortuitously with the arrival of an influential European geographer in the United States. Arnold Guyot, a disciple of the well-known German geographer, Karl Ritter, immigrated to America from Switzerland, carrying the Pestalozzian method of observational geography with which he had become familiar in Prussia.

As a result, the high schools of the nation began offering greater amounts of geography instruction (Stout 1921). In addition, there was increasing acceptance of physical geography as a scientific study (Stowers 1962). This curricular development coincided with the appearance of the first American physical geography text in 1855 (Rumble 1946).

Physical geography courses, especially in the secondary schools, had relied upon memorization and description. Guyot's philosophy for teaching physical geography was teleological, attempting to show that the great divisions of the world were not random or fortuitous occurrences but the work of the Creator. Mankind and nature and their interactions became the organizing structure for Guyot's geography. He was highly influential in increasing the commitment of secondary schools to physical geography in the curriculum, and his physical geography textbooks were readily and widely adopted (Fairbanks 1927). But his textbooks contained encyclopedic lists of material, forsaking his own belief that this was not the best way to teach geography.

Just as Pestalozzi had influenced Guyot, Guyot influenced William Morris Davis. Davis accepted the philosophy that geography should seek explanations and attempt to predict effects, consequences, and conditions. William Morris

Davis was to become one of the most influential geographers and educators in the history of American geography. He prepared textbooks and teachers' guides on teaching physical geography, including the activities of people in relationship to the natural environment (James and Martin 1979).

Stowers' summary of geography instruction in the United States from 1792 to 1892 concluded that the discipline matured significantly in content, scope, methods, and philosophy. Whereas geography content in the late 1700s consisted of vague descriptions of physical geography, by 1892 content included intricate explanations of the physical environment. Classroom instruction changed from recitation of factual knowledge with an indefinite status in the curriculum to a relatively permanent plan of study as an earth science (Stowers 1962).

Educational Reform Affects Geography

The year 1892 marked the beginning of change for the entire secondary school curriculum as a result of the report on secondary school social studies of the National Education Association's Committee of Ten (1894). The committee stressed that physical geography during the early secondary years, followed by physiography, should be recognized as the most important geographical offerings in the secondary school. The majority report of this committee receives credit for the policy decision on geography at the secondary level, thus resulting in physical geography's universal acceptance. However, Edwin Houston's minority report predicted that the newly recommended physical geography and physiography courses would be insufficient to meet the cultural demands of a changing society. He argued that the organic as well as the inorganic must be studied in terms of earth relationships, a rather unpopular theme at the time among geologists and physical geographers (Houston 1893). In 1897, the Science Section of the National Education Association (NEA) met to evaluate physical geography. It concluded that physical geography, with no human studies included, should remain the foremost offering in secondary school geography (NEA 1898). Davis attributed the increasing popularity of geography to the reports and publications of various National Education Association committees that resulted in thought-provoking discussions by teachers and administrators on geography in the curriculum (Davis 1902).

Regional Geography Emerges

In the early 1900s, a new trend emerged. Professional geographers began to favor regional geography as the most important geography offering in the secondary schools. In 1908 at the Association of American Geographers' meeting in Baltimore, Maryland, concerned professional geographers confronted the problem of high school geography and concluded that geography in the secondary school should "deal largely with regions—say, the United States and Western Europe" ("Geography for Secondary Schools" 1909).

A year later, the Committee of Seven of the Science Section of the National Education Association presented its report on geography and made the following recommendations (Chamberlain 1909):

1. Geography . . . should be, in some form, a required subject in all secondary schools. . . . 2. The subject should be pursued for not less than one year. . . .

3. The subject should be presented during the first year of the high-school course. . . .
4. There should be at least five recitation periods per week. . . . 5. About one-fourth to the total time should be devoted to the larger topics in physical geography, with the human side made more prominent than at present, and the remainder of the year [should] be given to a study of North America and Europe.

However, during the next several years, the general science curriculum assumed much of the content once assigned to physical geography. Teachers were not well trained in physiography, and topics were often dry, uninteresting, and removed from real life. Regional geography, on the other hand, was gaining a foothold in the curriculum, as well as commercial, industrial, and economic geography (Bengston 1929). Barnes indicated that the popular rise and then rapid decline of physical geography in the high school had permitted commercial geography to gain in prominence. Factual statements and statistics on the location, sale, and processes affecting the production of goods were the focus of commercial geography. It also included summaries of the production and trade of the major commodities of the world (Barnes 1934). However, it was through the regional approach that geography demonstrated its more humanistic traits in the curriculum with a people-and-place focus (Chamberlain 1909). Voluminous new textbooks began to employ a regional approach. They directed attention to the characteristics of various areas of the world with less emphasis on the products and trade approach of the older commercial geography courses.

Geographic Education and the Social Studies

Geography assumed a prominent role in the curriculum until the National Education Association (NEA) commissioned a secondary school curriculum review in 1911. This included an assessment of the significance of the social studies movement and its role in the curriculum. As a result, geography's role in the curriculum changed significantly. The NEA committee's report and recommendations placed the social dimensions of geography in the social studies (U.S. Bureau of Education 1916). The committee conceived social studies as representing a single field of study encompassing all the social sciences, without disciplinary boundaries.

The influence of the social studies on the elementary and secondary curriculum developed fully in the 1920s when a group of teachers undertook to design course materials. Most of the participants were historians and were ill-prepared to represent other disciplines in the social sciences. Recognizing this deficiency, the group requested assistance from subject-matter specialists in the other social science disciplines. All the social sciences responded except geography (James 1969). The prominent representatives of geography had determined that geography was not a social study and chose not to identify key concepts and ideas of the discipline despite an emerging professional literature suggesting geography's social dimensions (Barrows 1923).

As a consequence, geographers did not gain a prominent role in the early stages of the social studies movement. In fact, quite the contrary occurred. Nongeographers accepted responsibility for the geographic strand of the social studies curriculum. Those individuals often lacked training in and knowledge of geography and frequently made basic errors in information and map presentation. Rather than come to the rescue of the discipline, professional geographers withdrew but became increasingly critical of the poor examples of geography in the social studies.

For geographers, a partial resolution of the subject-matter disciplines/social studies dispute came in 1934. Isaiah Bowman, a distinguished professional geographer, reported on the American Historical Association's review of the social studies. Bowman (1934) submitted to the social studies and geographic professions the notion that geography was, at best, only a partially objective science since it included such variable phenomena as human societies. Preston James (1969, 479) summarized Bowman's message on social studies and geography.

> Geography, he insisted, must deal with processes, not with the mapping or describing of static things alone. He emphasized the importance of recapturing the thrill of discovery, which made the study of science exciting in the first place. The young people must be led to discover facts by deduction from theory and to formulate general concepts from the observation of apparently unrelated facts. This must be a major objective of teaching, he said, quite apart from an immediate social end that may be served.

Elementary School Social Studies Emerges

It was widely recognized as the social studies curriculum was developing that the discipline of geography, the child's view of the world, and classroom teaching intersected plainly in the elementary school. Selecting the elements of geography to include in the curriculum, identifying ways to highlight concepts and generalizations, and matching the content with the intellectual, social, and motor skills development of children each provided a complex maze of curricular design decisions. Social studies educators and the cyclical swings of educational philosophy influenced these decisions in the United States.

The Progressive Education Movement was a major influence during the early period of social studies curriculum development (Butts and Cremin 1953). This major force in elementary education during the first half of the 20th century promoted the concept of interdisciplinary social studies. As well as continuing the traditions of history and geography, the Progressive Education Movement placed greater emphasis upon the social sciences and developmental psychology than either history or geography. The major and probably the most durable outgrowth of the era for the social studies in the elementary curriculum was a philosophy that combined the natural interests of children with their increasing ability to comprehend more abstract elements of their expanding geographic and social environments. The curriculum came to be called the "expanding horizons" model and became the standard for elementary social studies in the United States (Hanna 1963).

The components of the expanding horizons curriculum were organized as concentric rings connected so as to have a spiraling, three-dimensional effect. It portrayed the effect of an increased interaction between the child and the environment as the child progressed through the grade levels in school. The innermost ring represented the child, the focus of study in kindergarten and the 1st grade. As the child moved into and through the early and upper elementary grades, the geographical and social scope of the content expanded to the neighborhood, community, region, or state, and by the 5th grade, the country, the 6th grade, the continents and countries of one of the hemispheres, and by the 7th grade, the world. Professional literature refers to this model as the expanding environment, journey geography, or age-mate geography.

The theory and philosophy underlying the expanding horizons curriculum are not without research verification. Children progress through a definite sequence of spatial competency, although the ages at which various children pass through the sequence vary. The research studies by Piaget and Weil (1951) and Stoltman (1977) have demonstrated the extent to which the development sequence of spatial stages generally followed by children complements the expanding horizons curriculum model.

Geography in Secondary School Social Studies Emerges

Changes continued to occur in the structure of the secondary school curriculum in the 1930s. Social studies either replaced or became a parent subject for industrial and commercial geography (Mayo 1965). The National Council for the Social Studies (NCSS) recommended that the "more mature aspects" of geography be taught at the high school level as part of the social studies. These included (1) "time, place, and space relationships," especially with reference to aviation, (2) the increased use of maps and the teaching of map skills, and (3) increased attention to "geographical factors and influence in economic, social, and political life in the past and the present, and in planning for the future" (NCSS 1944). The inclusion of geography in the social studies curriculum continued to progress rapidly. Special interest groups in geography comparable to those that had been concerned with the subject as a separate offering in 1892 and 1908 had little to say.

The fusion of geography with the social studies resulted in a new emphasis upon humans and regions in the geography segment of the curriculum. Some educators observed that the new interest in geography with a human point of view swept the nation (Lemaire 1946). The United States was involved in international endeavors. Vast oceans no longer permitted the nation to remain separated from the rest of the world. Geography in the modern secondary school was essential since all persons were, of necessity, students of world affairs.

Geography within Social Studies Education since the 1930s

As curriculum revision continued in social studies, the inclusion of social science content other than geography and history resulted in a multidisciplinary focus. Nevertheless, history and geography remained central as it proved difficult to achieve complementary foci or close coordination among the disciplines. Curriculum developers viewed geography with favor since it permitted the inclusion of the physical maze that people call home as well as the study of associated cultural manifestations that are related to particular geographical locations. The social studies continued to gain in importance as an essential element of the elementary school curriculum after 1935 with geography and United States history as its major components.

Geography in the secondary school curriculum has not fared nearly as well despite its strength in the elementary curriculum. Mayo (1965) summarized the situation:

(1) Physical geography has been neglected almost completely; (2) even in the relatively sparse offerings of human geography, factual material is left out in favor of the interest-centered appeal; (3) as a fused content subject, geography is not given equal time with history or other social sciences; (4) geography does not extend into the senior high school except rarely, and it is even left out in the junior high school to a large extent; (5) there is indifferent teaching in social studies concerning geography due to the lack of preparation of the teacher whose background is usually primarily in history courses; (6) the chronological approach inherent to history wins out over the spatial or geographical approach; (7) there is a lack of understanding on the part of school administrators and teachers as to what geography is and what it should encompass; (8) there is a lack of adequate class time in the curriculum for both history and geography on equal time levels.

Amidst that rather dismal state of affairs, there have been attempts to elevate geography's role in the curriculum. The High School Geography Project (HSGP) was unquestionably *the* significant curriculum development in secondary school geography during the past 50 years. Natoli's summary (1986) of geography in relation to curriculum movements provides insight regarding the interactions within the discipline, the inquiry approach to teaching and learning, and geographical education during the development of HSGP. However, Winston's summary (1986) of the same period reveals numerous problems faced by geography in the social studies, with the general conclusion that geography suffered a diminished role in the social studies for one main reason: there was clearly a confusion about the nature of geography among educators in general. Winston goes on to suggest that geography in the curriculum needs to be clearly and directly articulated.

The problem of a less-than-clear definition of geography as a subject has resulted in three major issues for geographical education. The inadequate definition of geography for curriculum purposes results in confusion by teachers. This has been addressed and largely corrected by the *Guidelines for Geographic Education* (Committee on Geographic Education 1984), which is addressed in chapters 1, 4, and 6. Second, academic geographers are not aware of the content needs of prospective teachers, so that introductory college level courses have not covered material appropriate to the elementary curriculum. Third, geographers have not taken the initiative in participating in national curriculum movements, which, like social studies, rely to a large extent upon geographic content and information. Those movements include environmental education, global education, and international education (Mehlinger et al. 1980). In each case, grass-roots participation by geographers is needed, but lacking. Geographers need not grasp for every curriculum idea that surfaces, but those with geographic content and perspectives require a participatory response. Lacking timely responses, geographers have been left largely as bystanders in several of the major curriculum movements of the late 1970s and the 1980s.

CONTEMPORARY GEOGRAPHY AND THE SOCIAL STUDIES

The history of the social studies curriculum is also one of continued debate about what it should be and how it should accomplish its task. Social scientists, professional educators, and the public all have strongly held notions—if often fuzzy and overlapping—of what the social studies should be. In the light of geography's historical development as a social study, it is necessary to summarize major traditions in the social studies, and describe how geography can fit into each of the traditions.

Defining the Social Studies

The mainstream rationale for the social studies rests on four enduring propositions. The propositions are not without problems or challengers, but they represent a relatively widespread and intuitively appealing basis for the social studies.

The purpose of the social studies is citizenship education. From Jeffersonian times, a foundation of our society has been the proposition that an educated population is crucial to a democracy. The core purpose of social studies in the nation's schools is thus citizenship education (Barr, Barth, and Shermis 1977). Cognitive development, accumulation of knowledge about history and geography, and the development of concepts and skills from the social sciences are means to responsible citizenship. Social studies education as citizenship education is a complex, changing, multifaceted goal. It will not be the same at all grade levels or with all students. In addition, part of the difficulty with defining citizenship education precisely is that it changes with society. Citizenship education today involves living in a global society, increased concern with people-environment relationships, and a greater emphasis on multicultural education than was the case 30 years ago. Part of the school's role is to respond to those changes. However it is defined or, perhaps, however many legitimate alternative definitions vie for a place in schools, citizenship education is still the most commonly agreed-upon general purpose for the social studies.

The heart of citizenship education is decision making. A society that places a high value on democratic government and on maximizing individual freedom within the law is a society that depends on the ability of a substantial portion of its population to make reasonable decisions. Although Shirley Engle's article (1960), "Decision Making: Heart of the Social Studies," has been debated and challenged, no single alternative definition has achieved such widespread acknowledgment. There are two problems with decision making as a goal. First, decision making is itself multifaceted, hard to define, and difficult to evaluate (Kurfman 1977). Second, even though the core of citizenship may be decision making, it is encased in a body of facts, concepts, skills, and values that do not always appear to be directly interrelated.

The social sciences are the primary basis of the social studies. Whereas a wide variety of disciplines can contribute to citizenship education, the primary bases of the social studies are certainly the traditional social science disciplines (history, geography, political science, economics, sociology, and anthropology).

Other disciplines (psychology), areas (ethnic studies), or movements (law-focused curriculum, global education, multicultural education) have made and will continue to make major contributions. Since there is clearly more valuable material than it is possible to teach, the key selection standard against which all content must be judged is whether this content will help develop citizens who can make informed decisions (Barr, Barth, and Shermis 1977).

Decision making involves knowledge, information-processing skills, and values. Most social studies experts agree that decision making is a complex mental process involving knowledge, information-processing skills, and values.

Within the social studies, there is, and will probably continue to be, substantial disagreement about what knowledge, which information-processing skills, and what kinds of values or value processes should deal with teaching decision making. The NCSS curriculum guidelines (1971) popularized participation as a fourth component. (Participation will not be discussed in this chapter because it belongs with the informal, rather than the formal, curriculum. Opportunities for participation will vary widely depending on student background, teacher interest, and community support.)

Barr, Barth, and Shermis (1977) have summarized into three traditions some of the major strands of the debate on the social studies. These are: the Citizenship Transmission Tradition, the Social Science Tradition, and the Reflective Inquiry Tradition. Figure 1 summarizes the different perspectives.

The traditions identified have been both a basis of conflict and a framework for agreement. Certainly, many academic social scientists tend to view the social science tradition as a key. Professional social studies educators, particularly in higher education, tend to espouse the reflective inquiry tradition, and much of the public adheres to the citizenship transmission theory. Each of those groups might also adhere to the belief that their tradition is the right and proper tradition, and should be valued over the others. The authors disagree, recognizing that there are advantages and disadvantages to each tradition.

The tradition of social studies as citizenship transmission is the easiest to clarify and teach and the easiest to explain to the public. The publicity generated by the spate of tests evaluating student awareness of place names in the world exemplifies the public perception of this tradition. Unfortunately, social studies as citizenship transmission also tend to focus on the least powerful aspects of social studies in terms of their abilities to provide creative insights to a range of situations. As a result, the citizenship transmission tradition can produce learning in the social studies at its most boring. In its worst form, it is vacuous "read the text" and "fill in the blanks" recall exercises. At its best, it provides students with a useful, factual, conceptual, and valuational base for dealing with the world.

The tradition of social studies as social science can provide a clear intellectual framework that is both powerful and transferable to a wide range of situations. It has strong advocates in higher education and has been the focus of some excellent curriculum materials. It also is the most complex approach to learn, partly because the social sciences that advocate use of academic structures as a framework for use in the schools only rarely incorporate the same framework into the introductory courses in higher education that form the content back-

Figure 1
THE THREE SOCIAL STUDIES TRADITIONS

	Social Studies Taught as Citizenship Transmission	Social Studies Taught as Social Science	Social Studies Taught as Reflective Inquiry
Purpose	Citizenship is best promoted by inculcating right values as a framework for making decisions.	Citizenship is best promoted by decision making based on mastery of social science concepts, processes, and problems.	Citizenship is best promoted through a process of inquiry in which knowledge is derived from what citizens need to know to make decisions and solve problems.
Method	Transmission: Transmission of concepts and values by such techniques as textbook, recitation, lecture, question and answer sessions, and structured problem-solving exercises.	Discovery: Each of the social sciences has its own method of gathering and verifying knowledge. Students should discover and apply the method that is appropriate to each social science.	Reflective Inquiry: Decision making is structured and disciplined through a reflective inquiry process that aims at identifying problems and responding to conflicts by means of testing insights.
Content	Content is selected by an authority interpreted by the teacher and has the function of illustrating values, beliefs, and attitudes.	Proper content is the structure, concepts, problems, and processes of both the separate and the integrated social science disciplines.	Analysis of individual citizens' values yields needs and interests that, in turn, form the basis for student self-selection of problems. Problems, therefore, constitute the content for reflection.

grounds of preservice teachers. Perhaps as a result, the tradition of social studies as social science is most dependent on relatively sophisticated, and thus relatively expensive, curriculum programs, such as those developed in the 1960s, and on fairly intensive in-service training. Last, the social-studies-as-social-science tradition establishes the least-clear link with citizenship education. At its best, it provides a context for the kind of cognitive development that can transfer to a wide variety of situations. At its worst, it is a self-serving tool of academicians trying to diffuse their discipline without a clear understanding of the practical and philosophical difference between elementary and secondary education and higher education.

The reflective inquiry tradition in the social studies may have strongest intellectual claim for being most likely to develop students with improved decision-making skills. By focusing on problems, and trying to replicate the kinds of situations students will encounter, reflective inquiry has the clearest and most direct approach to developing decision-making skills. In addition, by maximizing student involvement, the social studies as reflective inquiry provide the context most likely to excite student interest and generate enthusiasm for learning. In addition, the student-centered focus has an immediacy that is undeniable. Many social studies teachers believe that dealing with racial slurs in the classroom is more important than dealing with the history of slavery in the South and that a unit on drug abuse is more important that a unit on international trade, even though the latter is more defensible as part of a social studies scope and sequence. In addition, many of the teaching techniques (such as values clarification and role-playing) are relatively easy to understand and implement, and they use student experiences as content background, so that complicated data sets are not required. At its worst, however, poorly run reflective inquiry will degenerate into uninformed discussions that elevate personal bias and mindless brainstorming to the core of the social studies curriculum, leading both parents and children to ask "What's being learned here?"

Valuing Diversity in the Social Studies

We do not propose a single definition of the social studies, but recognize and value each of the three traditions. In practicality, there is no alternative. The social studies are, and will remain, diverse; those who deal with the social studies must accept reality.

We also accept and value diversity for other reasons. First, each of the traditions of the social studies makes a contribution to the education of children in a legitimate way. There are some things children should know, and some values must be taught. Concepts and skills in the social sciences can provide students with new and valuable insights into their present and future world. Investigating issues and problems that affect students directly is one way to develop problem-solving skills and formulate and clarify values. In short, elements of each of the traditions should be incorporated into the social studies scope and sequence in the K–12 curriculum. In addition, Jean Fair has suggested that there may be a developmental aspect to the three traditions, with the citizenship transmission tradition being the most common and perhaps the most appropriate in the elementary schools and the social science and reflective inquiry traditions being more important in middle and secondary schools (Barr, Barth, and Shermis 1977).

Second, different students have significantly different needs and backgrounds. A 7th grade class is not the uniform group that many curriculum outlines seem to assume. A social studies class in central Detroit may be considerably different from one in San Antonio, rural Iowa, or suburban Washington, D.C. Because student populations are different, the social studies should vary in response.

Third, and most important, teachers and other education professionals (not high-level academicians merely) are the people who should make decisions about the content and methods of the social studies. The diversity described above requires that professional educators make difficult decisions about a complex reality. The role of geographers is to teach geography. Academic geographers know and value geography. Geographers see their discipline making a valuable contribution to social studies curricula at a variety of grade levels based on a range of social science traditions. The role of the teacher is different. The professional educator's touchstone is not a subject, but children. The teacher's problem is not how to teach geography but how to teach children, not what content is most representative of geography but what content will help children most. We believe that teachers are the appropriate people to make curriculum decisions, and this will lead to an appropriate diversity in the social studies. The goal for academic geographers is to make the discipline more understandable, consistent, and more readily accessible so that teachers can improve their decisions about incorporating geography within the social studies.

Making Geography Accessible within the Social Studies

In recent years, progress in defining a social studies scope and sequence has been matched by progress in defining the content and focus for geography in the curriculum. The publication of *Guidelines for Geographic Education* (Committee on Geographic Education 1984) followed the publication of the NCSS preliminary scope and sequence (National Council for the Social Studies 1984), and, in general the two documents correspond closely (National Council for the Social Studies 1986). Both documents signal a critical convergence in the streams of thinking about geographic and social studies education.

The NCSS scope and sequence proposed the expanding environments framework for the elementary curriculum and the contracting environments for the secondary curriculum while making it clear that there is room for all three social studies traditions within a "Scope and Sequence." In addition, the Scope and Sequence specifies a strong role for geography, particularly in the upper elementary and middle school grades. *Guidelines for Geographic Education* defines the geographic content focus and skills that might be included in the expanding environment curriculum. In addition, *Guidelines* presents five fundamental themes in geography that form the framework for organizing geographic content and skills. (The themes are place, location, relationships within places, movement (spatial interaction), and regions.) Like the social studies, however, geography has been characterized by complementary and often competing traditions (Pattison 1964). As a discipline, geography may be viewed as a topic (knowing about world geography), as a social science (understanding the spatial organization of society), and as a point of view (using a geographic perspective). Those characteristics of the discipline correspond with the traditions of the social studies and incorporate the five fundamental themes from *Guidelines for Geographic Education* in different ways (Fig. 2).

Geography As a Topic—Studying Places

The expanding environments curriculum model tends to emphasize geography as a topic. The focus of studying geography as a topic is on learning about specific places (as in studying the geography of Mexico). The five themes constitute a framework for the content to be learned; the skills, values, and decision making tend to relate to specific places. With each unit, the teacher is attempting to have students understand where the place is located, what the place is like, how it is typical of other places (regions), why it is the way it is (interrelationships), and how it interacts with other places. Figure 3 presents a summary of this tradition.

Figure 2
GEOGRAPHY AND THE TRADITIONS OF THE SOCIAL STUDIES

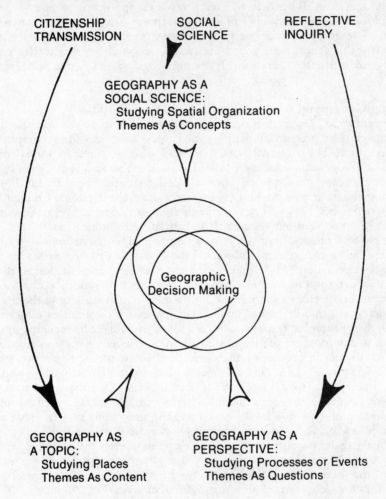

The skills associated with place geography are largely basic atlas and library research skills, and values issues revolve around identifying the values of the people of the community and one's own values with respect to other communities, including issues associated with ethnocentrism. The basic questions are: Would I like to live, work, play, in this place? Would I change, preserve, or avoid this place? How are the values of the people who live there similar to or different from mine?

The focus on specific places has both advantages and disadvantages. By focusing on a specific place, whether it is the child's community, or a com-

Figure 3
GEOGRAPHY AS A TOPIC—STUDYING PLACES

SOCIAL STUDIES TRADITION	Citizenship Transmission
KNOWLEDGE	Themes as Content
Location	Locations of places and regions
Place	Characteristics and importance of places
Interrelationships	Major human-environmental and cultural interrelationships of places
Interactions	Important interactions and movements
Region	Characteristics and importance of major world regions
INFORMATION-PROCESSING SKILLS	Find place locations
	Investigate places
	Compare and contrast places
	Use regions to generalize
VALUES	Like/dislike
	Values of different cultures
	Ethnocentrism
TYPICAL DECISIONS	In what kind of place would you like to live?
	What do you like/dislike about this place?
	How would you change this place? What would you preserve?
	What are the advantages/disadvantages of this place?
CLASSROOM ACTIVITIES	
Activities that focus on places include:	Models or maps of places, basic place location drills
	Collage of place characteristics
	Place location (and characteristics) drill (Where is the Amazon River Basin? What is it like there?)
	Atlas exercises (Find these places. What are the characteristics of this place? Where are there similar places?)
	Landscape interpretation (What do you see in this picture?)
	Compare and contrast places
	Travel games The connections of a place (Where do the things in this place come from? Go to?)
	Family migrations (Where did my family come from? Why did they leave their previous place? Why did they come here?)

munity or region about which the child is learning through media, pictures, maps, or text, the teacher can provide a concrete context for learning about the world. The disadvantage of studying places relates to motivation. Place geography can become a memorization exercise in learning unrelated facts that do not provide a basis for generalizing about the world and that have no immediate relevance to the learner. The key problem for teachers is motivation. Three ideas will help teachers teach about places and regions and maintain student interest.

Focusing is better than covering. Studying places does not mean that the teacher needs to do a regional inventory of the world. Most geography courses, at all levels, attempt to teach too much content. Focused case studies in different regional contexts will provide an easier framework for developing world awareness than attempting to cover the world superficially in a series of regional units.

Strive for relevance. Studying places is not an end in itself. Places provide a concrete context for studying the world, but their purpose is to learn more about how the world works as a background for citizenship. Part of the teacher's job is to continue to select geographic materials that are relevant to issues.

Use it or lose it. One of the important geographic content goals is to help students develop a place-name reference system upon which they can build. Unless the major place names are used repeatedly, they are forgotten rapidly. In addition, whereas an appropriate long-term goal of education is to help people develop a coherent image of the world, such an image continues to develop through the adult years as world events in different places impinge on people's lives. A short list of world place locations that students know well is better than a long list of place locations students learn superficially.

Geography As a Social Science—Studying Spatial Organization

Spatial organization may be defined as where things and people are located and how they interact. Geographers and others have developed concepts, generalizations, and principles to explain why things are located where they are, and how they influence one another. When teaching geography as a social science, the focus is on developing concepts and skills to help students understand why things are located where they are, and to decide where things should be located. Whereas the focus of geography as a topic deals with learning about specific places, the focus of geography as a social science deals with learning principles of location and interaction to help students understand all places. The five fundamental themes become organizing concepts and generalizations rather than content. For example, a 3d grade class might study the characteristics of its community as a geographic topic, and classify communities into urban, rural, and suburban as a means of developing geographic concepts. The same teacher might teach about the historical reasons for the home community's location and also teach generalizations about the location of cities (some cities locate near resources, cities often locate at transportation break points, cities locate in relation to other cities, cities function as service centers for the surrounding area). The skills students develop are associated largely with handling information as a context for location decisions. The values tend to be the criteria for the specific decisions (the least-cost location of a manufacturing plant, the location with greatest accessibility for a service concern, the quality of life vs. expense of a home location). Figure 4 summarizes major aspects of geography as a social science.

Figure 4

GEOGRAPHY AS A SOCIAL SCIENCE—STUDYING SPATIAL ORGANIZATION

SOCIAL STUDIES TRADITION	Social Science
KNOWLEDGE	Themes as Concepts
Location	Principles of location
Place	Classifications of places
Interrelationships	Principles about interrelationships
Interactions	Generalizations about movement
Region	Kinds of regions (uniform, formal, composite)
INFORMATION-PROCESSING SKILLS	Use regions as a means to classify information
	Collect, organize, and present data
	Analyze advantages and disadvantages of different locations
VALUES	Values and the criteria for location decisions
TYPICAL DECISIONS	Where should people, things, or activities be located?
	Why are people, things, or activities located where they are?
	What are the consequences of location decisions?
CLASSROOM ACTIVITIES	Analyze a location (Why is this community located where it is?)
	Choose a site (Where should we locate this industry, business, school, home?)
	Create regions (How can we divide this area into regions that make sense?)
	Analyze migration (What are the push and pull factors influencing migration?)
	Predict diffusion (Where will a good, service, phrase, song, style, attitude spread?)

The kinds of structured classroom exercises characterizing this type of geography lend themselves to small group activities, and are most useful in upper elementary and secondary classrooms. The range of decision making in structured location activities is particularly appropriate for students making the transition from concrete to abstract levels of thinking using geography.

Teaching geography as a social science has many advantages. The focus on making decisions about locations can provide an excellent context for developing general decision-making skills. In addition, the concepts and principles are readily transferable and generalizable to other situations. Unfortunately, textbooks do not provide particularly good materials, and the courses preservice teachers take in college often do not provide a good introduction to this tradition of geography. One strength of the High School Geography Project was that it included many exercises that teachers could use to teach geography as a social science.

37

Figure 5
GEOGRAPHY AS A PERSPECTIVE—STUDYING PROCESSES OR EVENTS

SOCIAL STUDIES TRADITION	Reflective Inquiry
KNOWLEDGE	Themes as Questions
Location	Where? (How is location of the event important? Where is the event with respect to other related places?)
Place	Who and what? (How do the human and environmental characteristics of places and regions affect the situation?)
Interactions	So what? (What are the consequences of the event for interrelationships and for other places through interactions?)
Interrelationships	Why? (How did interrelationships at the place, or interactions with other places, help cause the event?)
INFORMATION-PROCESSING SKILLS	Describe how the characteristics and location of a place influence an issue or event.
	Describe the causes of the event at the place and at other places.
	Describe the consequences of the event for people at the place and at other places.
VALUES	Consider the values of the different people involved in events.
	Relate individual values to the event or issue.
TYPICAL DECISIONS	What is a policy, solution, response, or personal position about an issue or event?
	What are the possible consequences of different policies or actions?
CLASSROOM ACTIVITIES	Create a map that tells a story.
	Write a newspaper article that uses a geographic perspective.
	Predict the consequences of an event for the place and other places.
	Write a dialogue about people from different places and cultures describing their perspectives on an event.

Geography As a Perspective—Studying Processes or Events

Geographers, business executives, feminists, historians, Marxists, psychologists, and many others have principles for organizing the way they view the world. Historians tend to see issues or events as developments in time (Fig. 5). They ask what is happening, why it happened at a particular time, what preceded and perhaps caused the events, what related things were happening at the same time as the event, and what its possible consequences will be in the future. Geographers ask *what* is happening, *where* it is happening, *why* is it happening *where* it is, how things at the place and at other places have helped cause it, and what the consequences for the place and for other places will be. We find many examples of using a geographic perspective outside the discipline. Many good news articles are contemporary geography in that they use the five fundamental themes of geography as questions. Who and what are involved (characteristics of places and the people who are there)? Where are they (location)? Why are they there (cause through interrelationships and interaction)? And so what (consequences for the place and for other places through interrelationships and interaction)? As geography teachers, we must be able to recognize this perspective. Teaching students to use a geographical perspective involves working with students to help them think about their world. It also incorporates understandings from the other traditions of geography. Students might learn about migration to the United States as a geographic topic, learn about a push–pull migration model as a geographic generalization, and use a geographic perspective to examine the process of migration from cities to suburbs. A student might study the characteristics of the home community, concepts about communities, and such issues as locating a solid-waste disposal facility within the community. In a single exercise, a student might learn about the border between the United States and Mexico, discuss the advantages and disadvantages of different kinds of national borders, and look at the issue of legal and illegal immigration.

Geography as a perspective is clearly within the reflective inquiry tradition of the social studies. Although the questions that ask who, what, where, why, and so what provide general guidelines for developing a geographic perspective, each person's perspective will be slightly different from every other's. As a result, teaching geography as a perspective links closely with developing the questioning abilities of students and makes use of such teaching techniques as inquiry teaching, values clarification, and individual assignments and reports that require the student to analyze, verify, and evaluate information. Writing, organization, and presentation skills, including developing maps that help present a story, are all involved. Addressing the values issue involves recognizing the differing values of the people being studied and the values of the writer. It also involves analyzing the values that shape public policy decisions and responses or solutions to public problems.

Teaching geography as a perspective is relatively straightforward, and introductory exercises are easy to design. Teachers can begin with a basic open-ended assignment. Teaching geography as a perspective requires an insightful and knowledgeable teacher, because critical thinking experiences for the students require the teacher to provide the right kinds of feedback to different students working on different problems at different levels of sophistication. Although a teacher can compartmentalize the teaching of geography as a topic or as a social science in a unit and feel confident that students accomplished some worthwhile short-term learning objectives, this is not the case with teaching geography as a perspective. Working on developing a perspective is a long-term, repetitive task. Essentially, learning to look at the world like a geographer (or like a historian or journalist) involves solving problems, receiving feedback, improving the solution, and working on more problems. In this respect, work on developing a geographic pespective will link well with teaching writing. The process of writing, receiving feedback, and rewriting is a method of examining and revising thinking that applies well to developing a geographic way of thinking.

REFERENCES

Barnes, Charles C. "The Place of Social Geography in the High School." *Journal of Geography* (May 1934): 178–86.

Barr, Robert D., James C. Barth, and S. Samuel Shermis. *Defining the Social Studies,* Bulletin No. 51. Washington, D.C.: National Council for the Social Studies, 1977.

Barrows, Harlan H. "Geography as Human Ecology." *Annals of the Association of American Geographers* 13 (1923).

Bengston, Nels A. "High School Geography—To Be or Not to Be." *Social Science and Mathematics* (October 1929): 693–701.

Bowman, Isaiah, *Geography in Relation to the Social Sciences.* New York: Charles Scribner, 1934.

Boyles, B. "History of Geography As a Subject in the Curriculum of the Elementary School from 1776 to 1880." Unpublished master's thesis, University of Chicago, 1926.

Butts, R., and Lawrence Cremin. *A History of Education in American Culture.* New York: Holt, 1953.

Chamberlain, James. "Report of the Committee on Secondary School Geography." In *NEA Journal of Proceedings and Addresses.* Topeka, Kans.: National Education Association, 1909.

Committee on Geographic Education, National Council for Geographic Education and Association of American Geographers. *Guidelines for Geographic Education: Elementary and Secondary Schools.* Washington, D.C., and Macomb, Ill.: AAG/NCGE, 1984.

Davis, William M., *The Progress of Geography in the Schools.* First Yearbook. Chicago: National Society for the Scientific Study of Education, University of Chicago Press, 1902.

Engle, Shirley H. "Decision Making: The Heart of the Social Studies." *Social Education* 24 (November 1960): 301–6.

Fairbanks, Harold W. *Real Geography and Its Place in the Schools.* San Francisco: Harr Wagner Publishing, 1927.

"Geography for Secondary Schools." *Journal of Geography* (February 1909): 121–25.

Goodlad, J., and H. Share, eds. *The Elementary School in the United States.* Chicago: University of Chicago Press, 1973.

Hanna, Paul. "Revising the Social Studies: What Is Needed?" *Social Education* 27 (April 1963): 190–96.

Houston, Edwin J., *Committee of Ten on Secondary School Studies.* Washington, D.C.: National Education Association, 1893.

James, Preston E. "Geography." In *The Social Studies and the Social Sciences,* ed. Bernard Berelson et al. American Council of Learned Societies and National Council for the Social Studies. New York: Harcourt, Brace and World, 1962, 42–87.

James, Preston E. "The Significance of Geography in American Education." *Journal of Geography* 68 (November 1969): 473–83.

James, Preston E., and Geoffrey J. Martin. *The Association of American Geographers: The First Seventy-Five Years, 1904–1979.* Washington, D.C.: Association of American Geographers, 1979, 16–17.

Kurfman, Dana, ed. *Developing Decision-Making Skills,* Washington, D.C.: National Council for the Social Studies, 1977.

Lemaire, Minnie E. "A Universal Subject—Geography for the Secondary School." *School Science and Mathematics* (April 1946): 366–68.

Mayo, William L. *The Development and Status of Secondary School Geography in the United States and Canada.* Ann Arbor, Mich.: University Publishers, 1965.

Mehlinger, Howard, et al. *Global Studies for American Schools.* Washington, D.C.: National Education Association, 1980.

National Council for the Social Studies. *The Social Studies Look Beyond the War.* Washington, D.C.: National Council for the Social Studies, 1944.

National Council for the Social Studies. *Social Studies Curriculum Guidelines.* Washington, D.C.: National Council for the Social Studies, 1971.

National Council for the Social Studies. Task Force on Scope and Sequence. "In Search of a Scope and Sequence for the Social Studies." *Social Education* 48 (April 1984): 249–62.

National Council for the Social Studies. "A Scope and Sequence Alternative for the Social Studies." *Social Education* 50 (November/December 1986): 484–542.

National Education Association. *Preliminary Report of the Subcommittee on Physical Geography.* Chicago: University of Chicago Press, 1898.

National Education Association. *Report of the Committee of Ten on Secondary School Studies.* New York: American Books Co., 1894.

Natoli, Salvatore J. "The Evolving Nature of Geography." In *Social Studies and Social Sciences: A Fifty-Year Perspective,* Bulletin No. 78, edited by Stanley P. Wronski and Donald H. Bragaw. Washington, D.C.: National Council for the Social Studies, 1986, 28–42.

Pattison, William D. "The Four Traditions of Geography." *Journal of Geography* 63 (May 1964): 211–16.

Piaget, Jean, and A. Weil. "The Development in Children of the Idea of Homeland and of Relations with Other Countries." *International Social Science Bulletin* 3 (1951): 561–78.

Rosen, Sidney. "A Short History of High School Geography." *Journal of Geography* (December 1957): 405–12.

Rumble, Heber E. "Early Geography Instruction in America." *The Social Studies* (October 1946): 266–68.

Stoltman, Joseph. "Children's Conception of Space and Territorial Relationships." *Social Education* 41 (February 1977): 142–45.

Stout, John E. "The Development of High School Curricula in the North Central States from 1860 to 1918." Unpublished doctoral dissertation. University of Chicago, 1921.

Stowers, Dewey. "Geography in American Schools, 1892–1935: Textbooks and Reports of National Committees." Unpublished doctoral dissertation. Duke University, 1962.

U.S. Bureau of Education. *The Social Studies in Secondary Education: Report of the Committee on the Social Studies,* Bulletin No. 28. Washington, D.C.: U.S. Bureau of Education, 1916.

Vuicich, George, and Joseph Stoltman. *Geography in Elementary and Secondary Education: Tradition to Opportunity.* Boulder, Colo.: ERIC/ChESS and Social Science Education Consortium, 1974.

Warntz, William. *Geography Now and Then.* New York: American Geographical Society, 1964.

Winston, Barbara. "Teaching and Learning in Geography." In *Social Studies and Social Sciences: A Fifty-Year Perspective,* Bulletin No. 78, edited by Stanley P. Wronski and Donald H. Bragaw. Washington, D.C.: National Council for the Social Studies, 1986, 43–58.

CHAPTER 4

GEOGRAPHY IN THE SOCIAL STUDIES SCOPE AND SEQUENCE

Joseph Stoltman and Michael Libbee

GEOGRAPHY IN THE SOCIAL STUDIES SCOPE AND SEQUENCE

The most widely recognized scope and sequence for social studies education in the United States is the 1983 Report of the Task Force on Scope and Sequence of the National Council for the Social Studies (NCSS 1984). Geography is not commonly taught as a separate subject in the early and upper elementary grades, but the NCSS Scope and Sequence incorporates geographic content and understandings within the curriculum. Discussion regarding the most appropriate scope and sequence for the social studies continues unabated (Bragaw 1986). We have chosen to use the 1983 Task Force Report as a guide to develop the following discussion to illustrate how geography can be strengthened within this format.

The authors have identified the major geographic education concepts, ideas, and topics from the NCSS Scope and Sequence Report for each grade level and summarized them briefly. The central focus for comparable grade levels as presented in *Guidelines for Geographic Education* (Committee on Geographic Education 1984) is then summarized. The summaries reveal how the two sets of curriculum recommendations complement each other. According to both sets, geography plays a significant role in the kindergarten through 7th grade curriculum for the social studies.

Several descriptions of learning opportunities for children at each grade level to enhance social studies and geographic education follow the summaries of the curriculum recommendations. The learning opportunities are taken from the comprehensive list of suggestions in *K–6 Geography: Themes, Key Ideas, and Learning Opportunities* (GENIP Committee on K–6 Geography 1987). The latter publication presents key ideas and recommended learning opportunities to accompany each of the five fundamental themes in geographic education: *Location, Place, Relationships within Places, Movement, and Regions.*

Kindergarten: Awareness of Self in a Social Setting. The scope of social studies during kindergarten builds upon and adds to children's knowledge about their social and geographical environment and the social life of the school, expanding upon the environment and social life of the home. In kindergarten, children learn about and develop the ability to move through the physical environment of the school and its vicinity. At this grade, children should begin to learn that they are individuals in a world of many different people and groups.

The **central focus is self in space.** Becoming familiar with and learning about places and objects in the environment within which children move about or with which they interact constitutes the major learning. The following are examples of essential learning opportunities for each of the fundamental themes.

Kindergarten children should have an opportunity to complete these or similar instructional activities for the following themes.

Location:
- Locate places in or near the school and describe them using such relative terms as 'next to', 'far', 'near'.

Place:
- Use pictures to describe the physical characteristics of places, such as home and school.

Relationships within Places:
- Give examples of the ways that people near the school depend upon the physical environment and how they have changed the environment.

Movement:
- Relate stories of how each child moves between home and school. Draw pictures of transportation that children use to travel between their homes and the school.

Regions:
- Draw or model the school environment using paper and crayons or a sandbox.

First Grade—The Individual in Primary Social Groups: School and Family Life. During 1st grade, the horizon of the social studies scope and sequence expands to include the individual in primary social groups. It is important to continue to develop ideas that demonstrate how spaces in the home and school are similar in some ways and different in others. The location of the home in a rural or urban place and in relationship to the school is an important reference point in learning. It places the ideas of earning a living and meeting material needs such as food, clothing, and shelter in the context of the home, school, and primary social groups. It uses the journey from home to work and back, from home to school and back, and from home to other places, such as shops and offices for medical treatment to reinforce the ways children expand their groups and environments with which they interact.

From a geographic education perspective, **the scope for the 1st grade is homes and schools in different places.** The operational environment of the first grader expands to include the relative location of the home and school. Children can explore streets and pathways that permit movement between home and school and become familiar with alternate routes to other places in the neighborhood. They examine the physical and human characteristics of both the home and school, and they also become familiar with the layout of school grounds and other nearby places such as parks and buildings. The following are examples of essential learning opportunities:

Location:
- Describe the location of home in relation to school.

Place:
- Use a map to locate and classify classmates' homes as near, far, closest to, or farthest from school.

Relationships within Places:
- Observe, describe, and record changes in the local environment over time.

Movement:
- Chart several ways to travel from one place to another.
- Map family migrations.

Regions:
- Use photographs to describe ways in which the local area has changed.

Grade Two—Meeting Basic Needs in Nearby Social Groups: The Neighborhood. The social studies curriculum scope at the 2d grade level centers on the concepts of production, consumption, communications, and transportation. Each of the concepts has geographic implications. Production entails collection of natural resources or component parts for assembly from a geographic region broader than the immediate neighborhood. Consumption entails meeting basic needs by relying upon environments beyond the local community. For example, oranges grown in climatically suitable places are consumed in many different places. Television and stereo sets, produced in several key places, are sold in many different places. People in different parts of the world find out about the availability of items or news about events through communications networks. Communications may originate down the street or on the other side of the world. The movement of component parts, agricultural and manufactured products, depends upon transportation. Exploring the role of transportation brings many different places on the earth closer together in travel time. Learning the significance of directions is important in questions of location. The physical characteristics of the landscape in the neighborhood represent a group that has different ties with people in other places, including other parts of the world. The 2d grade is a time to nurture the idea of a global perspective, comparing life as the children experience it with life in other cultures.

From the perspective of geographical education, **neighborhoods, small places in larger communities, are the central foci** with the opportunity to compare the neighborhood with communities beyond the local community. It is time to explore relationships developing between neighborhoods, dependencies developing among them, and examine how movement takes place within and among neighborhoods. The following are examples of essential learning opportunities:

Location:
- Use maps to identify the relative location of places outside the neighborhood.

Place:
- Identify mountains, hills, plains, islands, lakes, and rivers on maps and pictures.
- Map the neighborhood.

Relationships within Places:
- Compare ways in which students and others use the physical environment to meet their needs.

Movement:
- Identify places that people depend upon outside the neighborhood, such as farms, lakes, forests, cities, and indicate how these places are interrelated.

Regions:
- Compare ways in which the students' neighborhood is similar to and different from other neighborhoods in the community.

Grade Three—Sharing Earth Space with Others: The Community. The scope for the social studies in the 3d grade is the space the community offers people and the ways in which people share and jointly occupy that space. As with the 2d grade, the concepts of production, distribution, communications, and transportation have applications within geography. It presents the community as a specific place in a larger global setting, with connections of various kinds extending out to other places. It demonstrates how the local community depends upon other places for various items and how at times the community and other places are interdependent. It extends this concept from local to national and international levels.

From the perspective of geographical education, **the central focus at the 3d grade is the community: sharing space with others.** The profile of the physical and human characteristics of the community becomes the object of study. Ethnic diversity and the cultural dimensions of diversity in the community are important to develop when examining ways in which people share community space. It is important to study ways the local community interacts with other communities and places. Use the community to develop the concept of a region. Examples of essential learning opportunities follow:

Location:
- Point out on a map the location of the community, state/province, country, and continent relative to other places.

- Investigate the reasons for the community's location.

Place:
- Compare a map of the local and a nearby community to determine how attributes of the physical and human environment may have affected their locations, e.g., a river, highway, railway.
- Describe the characteristics of different kinds of communities (rural, suburban, urban).

Relationships within Places:
- Use photographs or newspaper articles to list ways changing technology has brought about large or small changes in the community.
- Investigate reasons for the locations of activities in the community.

Movement:
- Evaluate several ways available in the community for sending messages to other places.
- Investigate why people originally came to live in the community.

Regions:
- Locate the political boundaries for the community and identify the similarities that make the community a political region, e.g., one mayor and council, one taxation unit.

Grade Four—Human Life in Varied Environments: The Region. The expanding horizons sequence identifies the region as the content scope at the 4th grade level. The region, a fundamental concept in geographical education, usually entails the home state or another political unit smaller than the country. It is also feasible to introduce world geographic regions based upon physical, climatic, economic development (such as agriculture, industry), or cultural criteria. In some instances, one can select a combination of those criteria to identify and classify regions for study. Whatever the region selected for study, criteria for regional coherence usually emphasize the relationships between humans and the environments. Although it is possible to emphasize how humans adapt to the environment, the study should emphasize how human activities have altered the environment.

From the perspective of geographical education, **the central focus at grade four is the state/province, nation, and world.** The student develops an understanding of how to define larger regions and how to use the regional concept for identifying and classifying common characteristics of an area. Geographers identify regions of various sizes and characteristics by their major geographical features, i.e., physical and cultural attributes. A further learning objective is to compare and contrast the people living within regions, their ways of earning a living, and their physical environments. The following are examples of essential learning opportunities:

Location:
- Describe factors that influenced the locations of communities.

Place:
- Compare the human and cultural characteristics of urban and rural areas.

Relationships within Places:
- Examine a region of environmental stress and identify the causes and nature of the changes in the environment, e.g., tropical rain forests—overcutting timber; arctic tundra or ice fields—oil drilling and pipeline construction.

Movement:
- Display on a world map the birthplaces of relatives or the homes of the child's family, friends, or relatives.
- Describe the reasons for some human migrations.

Region:
- Demonstrate that by changing physical and cultural criteria one can alter the boundaries and shapes of regions.

Grade Five—People of the Americas: The United States and Its Neighbors. In the 5th grade, the social studies scope is on the United States and its close neighbors, Canada and Mexico. Emphasis is on studying the history and geography of all three countries. It examines the human geography of the United States, with its ethnic, cultural, and racial diversity, which is complemented by essential characteristics of its physical geography. It presents the physical and human geography of Canada and Mexico to deepen students' understanding of the countries that share the majority of the North American continent with the United States.

From the perspective of geographical education, **the central focus of the 5th grade is North America: United States, Canada, and Mexico.** Study will emphasize important ideas about comparing the places, the interactions occurring between them, and their physical and cultural regions. The following are examples of essential learning opportunities:

Location:
- Locate places relative to other places.
- Explain why different kinds of activities have different locations whereas different activities may compete for the same locations.

Place:
- Explain why landforms, climate, natural vegetation, resources, and historical events affect the distribution of population.

Relationships within Places:
- Explain how economic activities influence their locations.

Movement:
- Map the sources of products imported into the home country.

Regions:
- Compare and contrast physical and cultural regions within the United States.

Grade Six—People and Cultures: The Eastern Hemisphere. In the 6th grade, the social studies scope recommends covering the Eastern Hemisphere as a large region and examining the locations of its diverse cultural and political regions. The peoples and cultures focus gives a strong human perspective to regional study. Yet major landforms and water bodies will also have significance. The emphasis upon regional treatment necessitates classifying human and physical characteristics into coherent regions that are distinct from adjacent regions. The regional divisions should be made upon human and cultural elements such as language, political institutions, technological development, and belief systems. The regional economic development of the Eastern Hemisphere, including levels of development are topics of study. Natural resource bases, resource use, industrial development, and the extent of the service sector are all important considerations for studying economic development. Political geography may often be used as one means for classifying the region. A country's population, mineral resources, and level of economic development are important indicators of political influence. The interdependence between places in the larger region helps explain the development of significant patterns of trade, communications, government aid, and migration.

From the perspective of geographical education, **the central focus of the 6th grade is Latin America, Europe, the USSR, the Middle East, Asia, and Africa.** The knowledge and importance of location and relative location are key reasons for studying those regions along with general climatic patterns and landforms. The ways in which people have adapted to and used some harsh environments are highlighted. The movement of people along migration routes and the study of changing cultural regions over time are important to consider at this grade level. Critical elements to examine are the interdependence that develops between different regions of the world, the reasons for these global connections as well as the connections between developed and developing countries. A main focus is to identify major global issues with geographic dimensions. Among these are deforestation, pollution, and loss of farmland to desertification and erosion. From these, students can develop comparisons of the different ways people in different regions and from different cultures view issues with geographic dimensions. This will lead to examining and evaluating the relationship between personal choices and national choices of geographic importance and their global consequences. The following are examples of essential learning opportunities.

Location:
  • Evaluate a list of explanations for the location of a city.

Place:
  • Compare and contrast political, economic, and social characteristics of the region studied.

Relationships within Places:
  • Describe ways in which people adapt to and change the environment.
  • Discuss the advantages and disadvantages of different environmental changes.

Movement:
- Discuss the reasons for migration to the Western Hemisphere and within the Western Hemisphere.
- Describe the push-pull migration model.
- Interpret flow charts and maps that show direction and volume of trade among and between countries in different regions.
- Describe the reasons for different kinds of trade.

Region:
- Identify and evaluate criteria used to define developed or developing regions.

Grade Seven—A Changing World of Many Nations: A Global View. In the 7th grade, the social studies scope recommends covering the world in a global context. It develops the idea that the world is the home of people who are members of many varying culture groups. Those people try to deal with the large and small forces that affect their lives. The international content for this course will emphasize the major concepts in geography—resource distribution, spatial interaction, areal differentiation, and global interdependence. It should stress the aspirations of and changes and problems occurring in developing countries and it should emphasize the many cultural and economic interconnections that develop between places and people in the modern world.

From the perspective of geographical education, **the central focus of the 7th grade is state or regional geography, or world geography.** A course in state or regional geography enables students to study places at an intermediate scale, and to examine geographical problems of immediate, national, and global significance. It should provide opportunities to develop comparative analyses of areas within the state or its regional setting. Key geographic elements will include knowledge of places, their locations, patterns of climate, landforms, flora and fauna, population, settlement, transportation, and the economy. The world geography course should help students discern the global patterns of physical and cultural characteristics, such as earth/sun relationships, climate, population, transportation, economic interdependence, and cultural diffusion. Students should explore case studies of relationships between humans and the environment, such as those between rural populations and agriculture, or urban environments and industrial pollution. The following are examples of essential learning opportunities:

Location:
- Map the locations of places related to the global issues being studied.
- Describe the significance of location to a variety of events.

Place:
- Use geographic terms to describe the human and physical characteristics of places for writing assignments and oral presentations.

Relationships within Places:
- Evaluate the options available to people in different cultures as they rely upon the environment for basic natural resources.

Movement:
- Compare the feasibility of using transportation and communications linkages such as radio, television, and printed material to diffuse ideas from one region to another.
- Discuss how major migrations may have influenced the region.

Regions:
- Suggest and evaluate the ways governments in different regions of the world might cooperate to build a peaceful future for their people.

REFERENCES

Bragaw, Donald H., ed. Special Section, "Scope and Sequence: Alternatives for Social Studies." *Social Education* 50 (November/December 1986): 484–542.

Committee on Geographic Education, National Council for Geographic Education and Association of American Geographers. *Guidelines for Geographic Education: Elementary and Secondary Schools.* Washington, D.C., and Macomb, Ill.: AAG/NCGE, 1984.

Geographic Education National Implementation Project Committee on K–6 Geography. *K–6 Geography: Themes, Key Ideas, and Learning Opportunities.* Washington, D.C.: Geographic Education National Implementation Project, 1987.

National Council for the Social Studies. Task Force on Scope and Sequence. "In Search of a Scope and Sequence for the Social Studies." *Social Education* 48 (April 1984): 249–62.

CHAPTER 5

THE PREPARATION OF GEOGRAPHY TEACHERS

Dennis L. Spetz

"Suggestions for the Preparation of Elementary and Secondary School Geography Teachers" outlines the minimum competencies and the suggested course preparation in geography for (1) all who would teach geography in the elementary school and (2) all who would teach social studies or earth sciences in middle and secondary school.* Its potential audience includes institutions of higher education involved in teacher training, state teacher certification agencies, and national organizations for the accreditation of teacher education programs.

Essential Requirements for Preparing Teachers of Geography

The renewed interest in geographic education in elementary and secondary schools provided the Geographic Education National Implementation Project (GENIP) Committee with an opportunity to improve geographic instruction in a most significant way—that is, to suggest minimum requirements and essential elements in training geography teachers. This seems particularly fortunate for two reasons.

(1) Geographers can now assist in strengthening the discipline at a grassroots level by providing guidance for teacher-training and liberal arts institutions that prepare K–12 teachers. This is necessary if we are to overcome the serious problem of "geographic illiteracy" in our population ("Geographic Illiteracy Assailed" 1984; Grosvenor 1985). It is essential to start improving the background of elementary teachers because this will ultimately allow teachers in middle and secondary levels to build upon what students have learned earlier

---

*In October of 1985, a Geographic Education National Implementation Project (GENIP) Committee was charged with developing guidelines for teacher preparation in geography for geography, social studies, and earth science teachers in grades K–12. Committee members were:

Dorothy Drummond (St. Mary of the Woods College, Indiana); Arthur O. Forbes (Colorado Rocky Mountain School); Howard Johnson (Jacksonville (Alabama) State University); Fredric A. Ritter (Morgan State University, Maryland); Dennis L. Spetz, Chair (University of Louisville, Kentucky)

This committee met periodically from March 1986 until February 1987 to prepare a document entitled "Suggestions for the Preparation of Elementary and Secondary School Geography Teachers." The document was based upon *Guidelines for Geographic Education* (Committee on Geographic Education 1984).

The suggestions represent a general consensus and should not be taken as an official statement of any professional group. It is hoped that these suggestions will result in added systematic emphasis upon training qualified geography teachers.

and to avoid the annoying and time-consuming tasks of having to provide geographic remediation.

(2) Geographers must participate in the processes outlined by a recent proposal that may dramatically change the nature of teacher training in the United States. Concern about problems associated with the generally low quality of teacher training led deans at several major universities throughout the United States to establish a study group that examined ways to ameliorate or remedy these problems. This group, established in 1983 and known as The Holmes Group, was developed to discuss ways major research institutions could assist in enhancing teacher education (The Holmes Group 1986). A major goal of The Holmes Group is to improve the intellectual soundness of teacher education by providing future teachers with a solid background in academic subjects. The committee report makes it clear that

> competent teaching is a compound of three elements: subject matter knowledge, systematic knowledge of teaching, and reflective practical experience. The established professions have, over time, developed a body of specialized knowledge, codified and transmitted through professional education and clinical practice. For the occupation of teaching, a defensible claim for such special knowledge has emerged only recently. Efforts to reform the preparation of teachers and the profession of teaching must begin, therefore, with the serious work of articulating the knowledge base of the profession and developing the means by which it can be imparted. The Holmes Group recognizes the central importance of a strong liberal arts education in the preparation of teachers. Of all professions, teaching should be grounded on a strong core of knowledge because teaching is about the development and transmission of knowledge. With this in mind, The Holmes Group commits itself to phase out the undergraduate education major in member institutions and to develop in its place a graduate professional program in teacher education.

The GENIP teacher certification committee began its deliberations based upon these two factors, renewed public interest in geography and the proposals offered by the Holmes committee.

The committee initially deliberated two basic questions. First, what should students in grades K–12 know about geography? Second, what geographic concepts and skills should teacher preparation institutions be teaching teachers to answer the preceding question? To answer the first question, what students should know is contained in the detailed list of suggested learning outcomes in *Guidelines for Geographic Education*. In addition, the committee had access to a more detailed list of learning outcomes for elementary grades K–6 provided in a draft of the recently published document, *K–6 Geography: Themes, Key Ideas, and Learning Opportunities,* prepared for a GENIP committee chaired by Walter G. Kemball (GENIP Committee on K–6 Geography 1987). This document elaborated considerably on the geographic concepts and suggested learning outcomes for the K–6 level based on the five fundamental themes contained in *Guidelines*. The more difficult question was what we should teach to prospective teachers.

Developing suggestions for teacher training in geography at a variety of grade levels and covering several possible disciplines—i.e., social studies and earth sciences—presented special problems for the committee, because there appeared to be no single collection of courses that might serve such diverse populations. In addition, the realities of contemporary geography departments would have some teacher training programs serviced by a limited number of geography offerings in small institutions whereas large institutions might have a wide variety of course offerings in which prospective students could enroll. The committee agreed that prospective teachers should first be made aware of the unique position of geography in the curriculum by providing students with a unique spatial viewpoint. Also, teachers should learn about geography's relationship to other disciplines by asking the basic questions of *where* and *why there*. In addition, prospective teachers must perceive the role of geography as a bridging discipline between the natural sciences and the social sciences while understanding its unique spatial viewpoint. Finally, preparing teachers in geography should provide a background in the essential skills necessary to use maps, globes, and other descriptive material and also in the processes required to gather, record, and interpret geographic information from other means such as charts, graphs, and remotely sensed imagery.

A fundamental part of any teacher-training program, regardless of the grade level, must be a basic familiarity with the five fundamental themes in geography provided by *Guidelines*. Whenever possible, teacher-training programs should provide content material that relates to these fundamental themes and is applicable to meeting the suggested learning outcomes in the K–12 geography curriculum.

Minimum Geography Requirements for All Teachers

Several factors helped to shape the selection of courses that would provide a minimal training in geography. The first was to determine what geography classes might be most commonly offered in a majority of teacher-training institutions. A second concern was the nature of the syllabus for each course. That is, precisely what does each course cover? A final and more delicate concern involves teaching personnel. From personal experience, all members of the committee are aware of the irony that geography courses may often be so labeled in the college catalog, but in fact be taught by persons without either basic or advanced academic training in the discipline.

Given these caveats, the proposed minimum combination of courses takes two forms. The less desirable would be to require a two–semester sequence. This would include an introductory course in geography covering basic principles of cultural geography and physical geography. The second course would be a world-regional approach with some emphasis given to the location and place themes in addition to the regional concept and the fundamental application of *Guidelines* themes.

The more desirable of the minimum options would include three courses, not necessarily in any order. One would be a full semester of principles of physical geography with emphasis upon the relationships between humans and their environment. A second course would be principles of cultural geography. Regardless of the sequence of courses, every effort should be made to integrate the materials taught in the second course into materials learned in the first. Since many classes in grades 4–7 as well as most secondary level classes deal with world-regional geography, the third course would be a world-regional survey. This would build upon the physical and cultural geographic principles gained from previous courses and provide an emphasis upon locations and places. In order to maximize the benefits from these limited course offerings, college teachers of geography should use specific examples from the home state whenever possible. This will benefit the many teachers at the elementary level who are required to teach a course (or courses) dealing with their state and region. These examples can strengthen the role of geographic factors in a state's history and geography.

Although the committee suggests strongly this combination of courses for all teacher trainees, given the crowded curriculum already required of prospective teachers, the committee suggested that these courses could become part of the general education requirements. Whenever possible, these courses should be taught by personnel having no less than a master's degree in geography.

Specialized Elementary Courses

Geography should be an important part of the elementary curriculum. Most elementary programs begin with a focus upon the class and school, then gradually work toward the broader focus of neighborhood, community, state, and nation (see chapters 3 and 4). The concepts of location, place, human/environmental relations, relationships between places, and region and integral parts of these social and spatial units. In addition to those courses, the committee recommended that prospective elementary grade level teachers be given a social studies methods course. This course should incorporate the use of map and globe skills as well as abundant experience in using graphs, charts, atlases, and other teaching aids. In addition, the use of field trips including practice in observation and recording of data and how these can be used at appropriate grade levels is necessary to give reality to the geography they will learn and teach (Manson and Vuicich 1977).

Social Studies Teachers in Middle and Junior High School

As previously stated, geography classes in the upper elementary and middle grade levels often focus upon world regions. This world-regional view is frequently the focus of courses labeled "social studies." The committee accordingly recommended that prospective teachers take the following courses in addition to the two or three core courses previously mentioned: (1) a world-regional course with emphasis on the location and character of places (if this course has not been previously taken) and (2) a regional survey of the United States or North America. This course is absolutely necessary for teachers who plan to teach American history at the middle or junior high school level. If it is available, a course in historical geography of the United States would be an appropriate substitute, assuming that the prospective teacher had previously completed the required combination of physical and cultural courses and possibly the world-regional course. A third course could be selected from a group of systematic or topical courses. Because most students reside in or near urban areas or within a metropolitan area, a course in urban geography might be appropriate for the prospective teacher. As a complement to government courses that are often offered at this level, a course in political geography might also be suggested. Other suggestions include environmental, settlement, population, or economic geography. Regional suggestions include Middle or South America, Canada, or a regional course dealing with the Western Hemisphere or the developing world (Winston 1986).

Science Teachers in Middle and Junior High Schools

Geography is both a physical and social science. The physical aspects of geography can serve as important background and enrichment for earth science classes. Earth science has become a strong component of the middle and junior high school curriculum. Because of the need to strengthen this element, the committee believes that minimum preparation for this group of teachers would include the required geography courses previously listed, in addition to the following:

- A course in geomorphology that would include field work. The advantage of field experience is that teachers can often replicate it for middle and junior high school students.
- A course in climatology. Here again, teachers could develop appropriate activities for these levels of students. Possible activities could include using daily weather maps for tracking weather systems or constructing a weather station for local observations and forecasting.
- A course in map reading and air photo interpretation. The use of topographic maps in geomorphology exercises and the use of weather maps in climatology projects seem positive ways to provide students with direct experience with landform and weather phenomena. The constant use of maps, air photos, and remotely sensed imagery in print media would offer reasonably inexpensive and readily available teaching tools for augmenting science instruction.
- A course in instructional methods for teachers that incorporates using geographic skills required for teaching field observation and interpretation of the elements of the physical environment (Winston 1986).

High School Teachers in Social Studies Disciplines

It seems axiomatic that geography, with its unique spatial approach, should serve as a valuable bridging discipline with the social studies. Although the required courses for accreditation in each discipline offer few opportunities to include geography, teacher educators should consider strongly as either requirements or suggestions the following courses for teacher-training programs:

History. Teachers of history need basic geographic information—in particular, the locations and descriptions of places. Also, because most history courses are regional in approach, the prospective history teacher should have more than cursory understanding and application of the regional approach in geography. Therefore, history teachers should have a course in world-regional geography if it has not been taken previously as a part of the required geography program. A second course should involve the geography of the United States or, preferably, North America. Finally, a course in the historical geography of the United States or North America would allow history teachers the additional context of the geographic region as a stage and continuous setting for the development of historical events.

Government. Teachers in government or civics courses should have experience with political geography. Ideally, such a course would include local or regional examples, such as boundary disputes, regional changes and reasons for voting patterns, or political characteristics of selected populations. Training and experience in field surveys by prospective teachers offer a valuable opportunity for replication in the social studies classroom. A second course for teachers of government, when possible, should include experience with the regional geography of North America or the United States. This experience would help develop a sense of place and the regional characteristics that are integral for examining political and governmental questions.

Economics. Teachers of economics in secondary schools should have a course in economic geography in addition to the basic courses. The economic geography course should be structured to introduce the spatial aspects of the production, exchange, and consumption of items of value within and among contemporary societies. If the occasion warrants, an introduction to economic geography theory might be appropriate for advanced students. A second course for economics teachers should be a world-regional course if it has not been taken previously.

Sociology. Teachers of sociology could profit from a basic course in urban geography or population geography. A possible additional course would be the geography of North America or the United States and Canada, which normally contains specific materials dealing with demographic characteristics within the general regional context.

High School Geography Teachers

It has been noted previously that geography is often taught by persons with little or no formal training in the discipline. To teach geography effectively in the middle and secondary grades, teachers of geography should have the equivalent of at least a minor in geography. Optimally a major in the discipline would give the prospective teacher maximum flexibility in dealing with a variety of courses. Under no circumstances should the program of studies in geography be less demanding than that required of a liberal arts major or minor in the discipline. A portion of the minor or major for the prospective teacher should include a strong general education core. A broad range of liberal arts disciplines including work in humanities, mathematics, natural sciences, and social sciences is a necessary complement to the eclectic nature of geography. All high school geography teachers should have completed the three core courses recommended for all prospective teachers. This combination of physical and cultural elements plus the world-regional survey offers a solid foundation of experience for prospective teachers.

In addition to the core courses, teachers should have strong training in map reading and interpretation of air photos and remotely sensed imagery. The integral use of maps in geographic instruction is critical and all teachers of geography should not only have appropriate training but also feel comfortable in using maps creatively in the classroom.

Because of our regional setting, all geography teachers should have experience with a regional course in North America plus others, if possible. The teacher must become familiar with the regional concept and with its use as an instructional tool. In addition, teachers should have a minimum of two systematic physical and cultural courses to complement their training in the regional courses.

Finally, each prospective teacher should have specific training in instructional methods that use all the geographic skills they have been studying. These should include using maps, graphs, pictures, and other materials for presenting data that are necessary to investigate spatial phenomena, to interpret these findings, and communicate this information to others.

High School Earth Science Teachers

Standards for teachers of earth science proposed by the National Science Teachers Association emphasize the "interdisciplinary character of this field," and suggest that "prospective teachers should learn how to use the local environment as a laboratory (National Science Teachers Association 1984). In meeting these requirements, geography courses can play an important part in earth science teacher preparation. In addition to the three basic courses previously suggested, courses could include the following. First, a course in geomorphology, whether taught as geography or geology, to provide basic knowledge of earth-forming processes. Second, a course in climatology and meteorology for emphasizing the spatial significance of climatic and weather patterns. A third alternative would stress the interrelationships between humans and the environment in such a course as conservation of natural resources or resource management. In addition, courses in remote sensing, map reading, and air photo interpretation should provide valuable experience for presentation in the earth science classroom. If possible, a methods course incorporating geographic and other interdisciplinary skills would provide a fitting complement to the overall learning experience.

On the Training of Geography Teachers

The activities of the GENIP teacher certification committee were based upon a commonly held conviction that the training of teachers in geography could and should be improved. In the past, teacher training has often been less than systematic, producing teachers who are neither adequately trained in nor enthusiastic about their subject matter. Their students learned to dislike geography because it involved lecture materials of little interest spiced with large amounts of rote memorization of forgettable names of countless places, cities, products, and physical features. This is not geography as it ought to be taught nor is it the dynamic subject that geographers know. Above all, it is not the geography that will be taught by well-trained, enthusiastic teachers.

The training of teachers requires more than exposure to facts; it requires involvement with the theoretical bases that place factual materials in meaningful contexts. Although using facts provides information students can use to understand the major themes and concepts of geography, the focus of teacher training must include conceptual materials such as those provided in *Guidelines*. Here geography can be presented as a unified field of study asking the basic questions *Where?, Why there?, What is its significance?,* and even questions of the optimal location of things. In seeking answers to those questions, students can have a rich and rewarding experience of geographic inquiry by observing geographic problems, gathering data, developing hypotheses, and explaining the nature of spatially variable phenomena.

Finally, those who teach prospective teachers should be aware, as Edward Fenton (1966) has observed, that "most students learn to teach by imitating their teachers." If geography is to grow and flourish and use current public interest to improve its status among the disciplines, teacher trainers in geography must give attention not only to geography's message but also to the manner in which it is presented.

REFERENCES

Committee on Geographic Education. National Council for Geographic Education and Association of American Geographers. *Guidelines for Geographic Education: Elementary and Secondary Schools.* Washington, D.C., and Macomb, Ill.: Association of American Geographers and National Council for Geographic Education, 1984.

Fenton, Edward. *Teaching the New Social Studies in the Secondary Schools.* New York: Holt, Rinehart and Winston, 1966, 2.

GENIP Committee on K–6 Geography. *K–6 Geography: Themes, Key Ideas, and Learning Opportunities.* Washington, D.C.: Geographic Education National Implementation Project, 1987.

"Geographic Illiteracy Assailed." *New York Times,* December 14, 1984, A24.

Grosvenor, Gilbert M. "Geographic Ignorance: Time for a Turnaround." *National Geographic Magazine* 167 (June 1985): editorial page.

The Holmes Group, Inc. *Tomorrow's Teachers: A Report of the Holmes Group.* East Lansing: The Holmes Group, Inc., 1986.

Manson, Gary, and George Vuicich. *Toward Geographic Literacy in the Elementary School.* Boulder: Social Science Education Consortium, Inc., 1977.

National Science Teachers Association. "Standards for the Preparation and Certification of Secondary School Teachers of Science." Washington, D.C.: National Science Teachers Association, 1984.

Winston, Barbara J. "Teaching and Learning in Geography." In *Social Studies and the Social Sciences: A Fifty-Year Perspective.* Bulletin No. 78, edited by Stanley P. Wronski and Donald H. Bragaw. Washington, D.C.: National Council for the Social Studies, 1986, 43–58.

CHAPTER 6

THE FUNDAMENTAL THEMES IN PRACTICE:
NUCLEAR EXPLOSION AT CHERNOBYL—AN EXAMPLE OF GLOBAL INTERDEPENDENCE

Robert W. Morrill, James Sellers, and
Stephen A. Justham

In the final analysis it came down to which way the wind was blowing.—
Dr. Robert Peter Gale.

Teachers should recognize that many contemporary events have enduring social and environmental significance and have a place in a school's social studies program (chapters 3, 4). Which events to select is problematic. Which themes and teaching methods to adopt is debatable. Further, it is natural that the choice of perspectives to use in examining contemporary events will nearly always be based on personal preferences and grounded in one's own subject-matter expertise. One essential perspective is offered in the recently published *Guidelines for Geographic Education* (Committee on Geographic Education 1984):

> Many . . . serious or potentially serious problems have geographical implications. Geographic knowledge is crucial in dealing with issues such as nuclear armaments buildups, siting nuclear power plants, safe disposal of radioactive and toxic chemicals, segregation by race, age, or economic status, discrimination against women and minorities, and inequitable distribution of economic resources in and among developed and developing countries. Careful geographic scrutiny can benefit the analysis of problems of environmental degradation, rational use of ocean resources, the resettlement of refugees from war-torn nations, and political repression and terrorism. Major policies formulated by governments and large corporations have impacts that reach around the globe.

Interest among teachers in contemporary events and world issues is not new. In 1980, John Wise (1980) wrote that for the preceding 90 years "geographic education literature has been continuously sprinkled with recommendations for the geography teacher and student to be especially aware of current events and global issues."

In trying to determine what parts of the world and which events to select for study, Wise compiled a list of 25 cities and 50 topics of the future that he expected to be prominent to the news media for two decades or more. Most teachers of social studies no doubt have lists of topics they consider of enduring

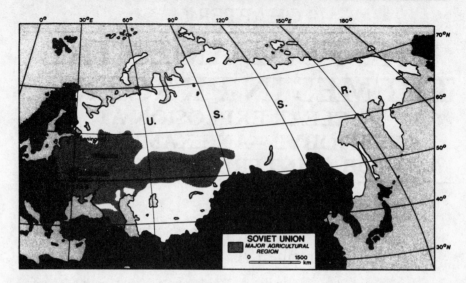

Figure 1

interest and relevance. Some teachers in 1980 may have listed the hazards of nuclear power plants. Surprisingly, Wise did not, although he included nuclear arms testing. On the other hand, it is very unlikely that in 1980 Chernobyl would have been on any teacher's list of important places. Chernobyl literally exploded into international prominence as a place and as a symbol of the hazards of modern technology.

An event that occurs in a short time and at a great distance from us can have a profound and lasting influence on our lives. The effect of a catastrophic event can vary in time, in extent of area affected, and in the number of people involved. As anticipated, the impact of radioactive contamination from the nuclear accident at Chernobyl was most intense in the immediate vicinity of the nuclear reactor. The contamination did not spread equal distances in all directions from the location of the accident (Fig. 1). As with other air- and water-borne contaminants, the spread of the radioactive elements was influenced by wind patterns and patterns of precipitation. The winds carried the contaminants across the USSR and beyond many international boundaries, as well as over large water bodies. The earth's physical processes took over the spread of radioactive material and respected no political and cultural boundaries (Fig. 2).

The Chernobyl explosion and its aftermath provide rich and diverse material for study in social studies classrooms. Although the initial event occurred in a seemingly obscure Ukrainian town, the consequences are global in scale, persist over time, continue to influence a large portion of the world's population, and will require international cooperation to understand and resolve. Scale, time, number of people involved, and level of international cooperation required for resolution are reasonable criteria to apply in selecting a contemporary event to include in the social studies curriculum. The explosion at Chernobyl on April

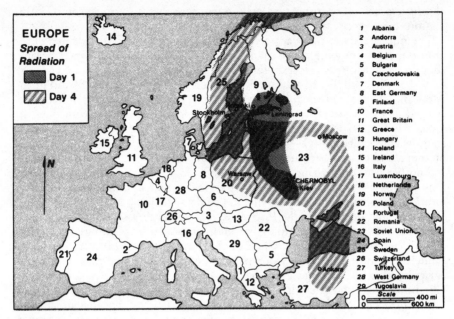

Figure 2

26, 1986, attracted worldwide attention and concern and triggered events that will influence the peoples of the earth far into the future.

Significant contemporary events are multifaceted and complex. Without an orderly framework of ideas for examining events, we risk being overwhelmed with unrelated bits of information and a jumble of conflicting interpretations.

By using the five fundamental themes of geography as developed in *Guidelines for Geographic Education,* we can enhance our understanding of complex events. In this chapter, we apply the five fundamental themes in geography to the Chernobyl accident. The interrelated geographic themes of *location, place, relationships within places, movements,* and *regions* provide a manageable set of concepts that teachers can use to organize and analyze the information related to a contemporary event. The chapter will conclude with suggested teaching activities that incorporate one or more of the five themes.

Chernobyl as a Location: Position on the Earth's Surface

Events, whether large or small, significant or insignificant, destructive or beneficial, all occur at specific times and locations. Human actions and physical environmental events do not occur in a void. Where things happen is crucial and is an inseparable aspect of events themselves.

It is helpful to use the arbitrary grid system of latitude and longitude for measurement and precision when locating places. Latitude and longitude give us numerical coordinates for measuring distances and finding directions among places on the earth. Although numerical coordinates that give an exact or *absolute location* are important, it is often necessary to acquire more information to appreciate the significance of a place's location.

The nuclear power plant accident at Chernobyl provides an example of the importance of knowing more than a place's absolute location. Upon hearing about the Chernobyl explosion, few persons would have been satisfied to learn only that Chernobyl's latitude and longitude are 51°17′N and 30°14′E (Fig. 1). After all, Chernobyl is near the site of a nuclear reactor installation that experienced an explosion, fires, and a partial meltdown, thereby releasing deadly radioactivity into the environment. Because of the explosion, people were anxious about Chernobyl—where it is located, what it is near, what is going on there, and how events there related to other places.

Another way to describe the location of a place is to use information about how the place's location is related to other locations—in other words, *relative location.* For example, Chernobyl could be described as being a small town of 12,000 people situated on the Pripyat River, which flows into a reservoir on the Dnieper River and is 600 miles (968 kilometers) south of Leningrad. Another description may mention that Chernobyl is 9 miles (14.5 km) from the site of a nuclear facility containing four nuclear reactors and is surrounded by an agricultural region of rye growing and dairying (Fig. 1). An atlas reveals Chernobyl surrounded by the symbols indicating a large area of swampy land, which may indicate further that the land is quite flat. Yet another description may state that Chernobyl's reactors are part of a large network of power-producing facilities throughout the Soviet Union and that the loss of Chernobyl's reactors meant power shortages for places as far north as Estonia on the Gulf of Finland. Still another description might include that Chernobyl is a small town located in the Ukraine—the breadbasket of the Soviet Union—432 miles (691 km) southwest of Moscow, and 80 miles (128 km) north of Kiev (Fig. 1).

After the explosion, knowing the relative location of Chernobyl became important to people in all areas on the earth. Many people wondered how far Chernobyl was from their homes, whether the nuclear accident would affect them, and how soon radioactive fallout might arrive where they were living.

The events at Chernobyl affirm that *locations,* regardless of how seemingly isolated or remote, are interconnected in webs of relationships. Distant happenings may influence our daily lives whether or not we want them to and with or without our knowledge.

Chernobyl as a Place: Physical and Human Characteristics

Places have physical characteristics, such as vegetation, soils, climate, and landforms. Physical characteristics of places usually are the result of long-term earth processes that give rise to specific arrangements of water bodies, topography, and animal and plant communities. Places are shaped and given meaning through the ideas and actions of people. Places can be distinguished from each other by the ways in which they have been given character by people from different cultures. The same place may have different meanings for different people and at different times. Also a place can take on special meaning because of events that occur there. A place can become a symbol.

The meaning of Pearl Harbor was transformed by a devastating surprise military attack. The place names 'Hiroshima' and 'Nagasaki' are forever identified with the horror of nuclear war. Auschwitz, Buchenwald, and Dachau are symbols of the senseless mass extermination of innocent people. Bhopal stands for the lethal dangers present in the production of modern chemicals. Chernobyl now symbolizes the devastation and far-reaching consequences of mishandling nuclear energy.

Before the nuclear explosion, Chernobyl was a small town filled with people pursuing normal daily activities of working at various jobs, managing households, having babies, raising families, growing gardens, spending time with relatives and friends, going to school, caring for the sick, and burying the dead. For the people living there, Chernobyl was home—a *place* where they wanted to be. All the activities and decisions of the people gave a distinctive meaning to Chernobyl. The homes, workplaces, streets, recreation areas, farmland, hills, rivers, forests, and the nearby nuclear power facility were all woven together and parts of a distinctive landscape. The activities of the people, combined with the human-made features and natural features of the area, made Chernobyl what it was—a place in which they had roots.

Relationships within Places: Humans and Environments

The Ukraine conjures visions of Cossacks with their swords raised high, rumbling across rolling fields of flowing grass as they charge to meet the rampaging hordes attempting to conquer them for control of the rich land of the Ukrainian steppe. Long the target of its neighbors, the Ukraine has been most commonly described as the breadbasket of the Soviet Union (Edwards 1987). The seemingly endless rolling steppe plays host to a variety of agricultural products in addition to the wheat associated with the 'breadbasket' label. Among the other leading agricultural products of the region are potatoes, sugar beets, corn, sunflowers, cattle, hogs, dairy products, and a variety of other crops.

Vast deposits of coal and iron have resulted in the Ukraine's industrial prominence. Steel, machinery, chemicals, and transportation equipment, including automobiles, farm machines, ships, and aircraft, are manufactured at various locations throughout the state. Agricultural and industrial productivity has created a number of large industrial centers. The Ukraine has five such cities with populations in excess of a million: Kiev, the capital in the north central part of the state with nearly 2.5 million; Kharkov in the northeast, 1.5 million; Dnepropetrovsk, in east central Ukraine, slightly more than 1.1 million; Odessa, perhaps best known as a resort on the Black Sea, home to more than 1.1 million; and Donetsk, in the southeastern Ukraine with just over one million. The Ukrainian S.S.R. is about 90 percent the size of Texas but has more than three times its population. The Ukraine contains only 2.7 percent of the land area of the Soviet Union, but produces anywhere from 20 to 35 percent of the nation's various agricultural and industrial products. Such activity demands large amounts of energy to sustain it. Although the Soviet Union and the Ukraine in particular apparently have surplus quantities of fossil fuels, the central government has turned to the inexhaustible energy of nuclear power plants to supply an increasingly larger percentage of its power requirements. Chernobyl, with four reactors on-line and two more in the construction stage, was designed to be the largest production capacity nuclear power plant in the Soviet Union.

Soviet central government concern for the environment has historically been of only passing interest, especially when it conflicted with national priorities and particularly when the issue was one of economics. Hills of mine waste abound in the anthracite, iron, and titanium mining regions. Growing mounds of furnace slag dot the iron and steel manufacturing areas, and dark clouds of smoke emanate from oil-refining plants. These environmental problems have often been ignored, because they are either localized or dissipate invisibly into the atmosphere in a relatively short period of time. The immediate environmental and human consequences of the nuclear disaster at Chernobyl, dramatic and headline-grabbing though it was, paled in comparison with the long-range effects on the environment.

The destruction of the Number 4 nuclear reactor at Chernobyl led to radioactive contamination of a land area of about one thousand square kilometers or 620 miles around the power station, an area about half the size of the state of Rhode Island. Whereas the atmospheric contamination that occurred dispersed and was carried across international boundaries, creating protests, effects upon the immediate environment will be experienced for decades, if not for generations. Nearby farmland will be unusable for the foreseeable future. Radiation contamination is not a one- or two-dimensional problem; it is omnidimensional. That is, it was emitted *into* the earth as well as on its surface and into the atmosphere and spread over a large area as a result. Therefore, immediate and continued contamination of the ground water supply and, ultimately, of surface waters within the water basin of the Pripyat River, Kiev (Kiyev) Reservoir, and the Dnieper (Dnepr) River must be a prime consideration of the Soviet government (Fig. 3).

This obvious lack of concern for humans and the environment, as well as their interconnectedness, has had severely adverse results. Governmental attitudes toward any aspect of the human-environmental condition can usually be masked easily or hidden from view under the Soviet form of government. However, no artificial boundary can prevent a disaster of the magnitude of Chernobyl from being known worldwide. Chernobyl also brought to the world's attention the fact that the Soviets had forsaken accepted safety construction practices in order to reduce the economic costs involved in such projects. Thus, Chernobyl is a dramatic example of placing environment at risk for the sake of cutting project costs. The Soviet government modified and adapted the Chernobyl environment to suit its needs. In doing so, it revealed certain perceived cultural values, economic and political circumstances, and technological misunderstanding of the potential dangers of the nuclear power generation industry. Although the Three-Mile Island accident in Pennsylvania hinted at the potential of a nuclear disaster, Chernobyl demonstrated explicitly the effect nuclear disasters may have upon both immediate and distant environments and the earth's atmosphere.

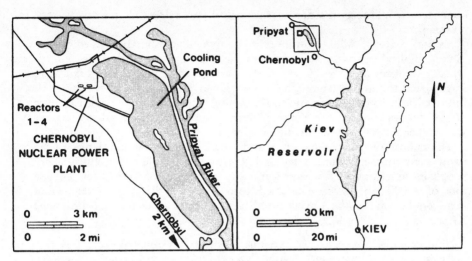

Figure 3

Chernobyl and Movement: Humans Interacting on the Earth

In the natural environment, *movements* on the earth's surface may be as swift as the whirling winds of a tornado or as slow as the wearing down of a mountain range. Some movements in physical processes are fast and brief; others are slow and take millions of years to complete. The speed and range of human actions have increased dramatically in the modern era—taking us beyond the earth's surface in transportation, communications, and aspirations.

Movements of people, ideas, products, and various aspects of the natural environment constitute an essential part of the continuous changes occurring on the earth. Movements brought about by human activity or natural physical processes have significant geographical and societal consequences.

In the midst of day-to-day social activities and human interactions with the natural environment, unpredicted, dramatic, and even tragic events may burst into our lives. The nuclear catastrophe at Chernobyl set in motion widespread human responses to the dangers of radioactivity. Clouds moving across the steppes of the Ukraine, onto the eastern European plain, and into Scandinavia dispersed their deadly load of radioactive contaminants. Winds and rains bearing radioactive elements gave new urgency to our need to comprehend this century's increasing interconnectedness of peoples and places. People the world over riveted their attention on the movements of the radioactive clouds stretching from the crippled power plant. Feeling vulnerable and exposed, government officials and citizens of many lands wondered aloud where the radioactivity would go. Many people quickly realized that "how far it would travel and whom it would affect depended on the vagaries of meteorological patterns" (Greenwald 1986, 39–52). The inseparability of the natural environment and human life was revealed anew as people struggled to reduce the negative consequences of the airborne radioactivity. Patterns of climate and unpredictable changes in daily weather became the arbiters for which parts of the earth's

65

ecosystem would receive a dose of humanity's most dreaded and deadly contaminant.

As with a line of falling dominoes, one movement inevitably led to others. "Like a biblical calamity, the impact of the accident seemed to be felt everywhere" (Greenwald 1986, 44–46). The intensity of concern was highest in Europe and the immediately affected areas, but people throughout the earth followed hourly reports on radio and television. The early and expected absence of Soviet news coverage of events in Chernobyl convinced people that whatever was happening was important and newsworthy.

The radioactive contamination in and around Chernobyl had damaged the natural environment, human settlements, and the people themselves. Many people in the vicinity fled as they learned of the dangers. Within a few days, a total of 84,000 people were evacuated from the area of the nuclear plant—first in a six-mile zone and later in an area that extended to an 18-mile (30-km) zone (Greenwald 1986, 44–46).

> Ultimately 135,000 people in an 18.6 mile [30 km] radius around the reactor were relocated; those relocated included the citizens of the town of Chernobyl, located nine miles [14.5 km] away from the plant, who found themselves relocated as late as one week after the accident began. As the government further understood the gravity of the situation, it sent several hundred thousand children from Kiev [80 miles (128 km) to the south], Byelorussia and northern Ukraine to recreation camps (Ramberg 1986–1987).

The effects of the Chernobyl explosion spread outward causing severe environmental damage, and precipitated the displacement of hundreds of thousands of people within the Soviet Union (Fig. 2, Day 4). Across Europe, tourists, foreign students, and guest workers fled many areas of Europe to avoid being in the path of falling radioactivity. The immediate effects of the disaster were most damaging to the people within the Soviet Union. Some died of radiation; others lost farms, homes, and jobs. Some would later lose status and membership in the Communist Party and be forced to endure criminal trials for their roles in causing the disaster.

Discussion, protest, and debate about political, economic, social, environmental, and medical problems reverberated across the globe. Almost no one would be left untouched and undisturbed by the clouds and winds that emanated from the Chernobyl area.

The human, environmental, and economic costs are difficult to estimate. "Along with the human costs are still uncertain economic ones. The Soviets place the direct cost at three billion dollars" (Ramberg 1986–1987). The accumulated costs of radioactivity in water and food supplies, the loss of productive agricultural land, the long-term cost of health care, and the impact on energy resource decisions throughout the world are probably impossible to calculate.

"It is impossible to hide things that betray their own existence [such as] nuclear disasters. Once out of control, atoms cannot be restrained and kept in one's country. They cannot be prevented from escaping over the frontiers, regardless of how fortified and protected they are" ("Eastern Europe and Chernobyl . . ." 1986). During the two weeks immediately following the meltdown, the winds blew northwest from the Ukraine. Not only were areas in northern Europe (Finland, Sweden, and Norway) affected; in the opposite direction, the Balkan nations of southern Europe (Bulgaria, Romania, and Greece) were also mildly contaminated. "Radiation spread across Europe from the start. Days after the accident, prevailing upper-level winds swept radiation over the Arabian Peninsula, Siberia, and eventually North America (Ramberg 1986–1987, 311).

Countries near the Ukraine acted to protect their citizens and reacted in diverse ways to the incident. Poland and Romania distributed potassium iodide tablets to children to minimize damage to thyroid glands. In Poland, milk and vegetables were dumped fearing that they had been contaminated. Yugoslavia put its entire nuclear power program on hold until the extent of damage and causes of Chernobyl could be more closely analyzed. Bulgaria suffered a 30 percent decline in Western tourists visiting its Black Sea ports in the summer of 1986. Officials decided that future nuclear power plants in Czechoslovakia would have protective shells of reinforced concrete. In East Germany, church leaders and political dissidents questioned their country's nuclear policies. Romania announced that it was going to review its nuclear energy program. Hungary made payments to compensate local agricultural producers and distributors for their losses. Exports were returned to many Eastern European nations when importing nations refused to accept what they feared were contaminated goods. Livestock losses to Western European nations were estimated at over 25 million dollars ("Chernobyl and Eastern Europe . . ." 1987).

The disaster at Chernobyl and the human and environmental consequences of the movement of the radioactivity are a clear example of interacting and interlocking movements on the earth's surface. Human and physical systems can be damaged and disrupted by a catastrophic event, an accident, a release of contaminants into the environment. Humans interacting through trade, transportation, and communication are normal parts of day-to-day activities and we take for granted such movements and interconnectedness. As we shape our environments and landscapes, we must remember that one accident can alter our relationships to parts of the earth for many years to come.

Chernobyl: A Part of Several Regions

Region is one of geography's most widely used concepts; it is "the basic unit of geographic study" (Committee on Geographic Education 1984). A region is an area characterized by distinguishing and unifying features that is distinct from surrounding areas.

Regions may be small or large, well known or obscure, old or new. Regions may be identified by physical characteristics, human-made characteristics, or by combinations of the two. As with other aspects of the earth, regions undergo formation and change and may even go out of existence.

Regions abound on the earth's surface. Well-known regions include the central United States' corn and soybean agricultural areas, impoverished but mineral-rich Appalachia, the alpine mountains of Europe, the Islamic nations of the Middle East, the rain forests of central Africa, French-speaking Quebec, and scores of political states that attempt to unify internal political differences within secure boundaries.

Regions, therefore, may be comprised of economic characteristics such as agriculture, industry, or retailing, or they may encompass natural environments such as a mountain range, a rain forest, or a swampland. Cultural characteristics, including language, religion, food habits, or voting behavior, may be used to distinguish regions. Similarities in political beliefs and practices may lead groups of nations to form multinational political regions.

Clearly, the geographical term 'region' is dynamic, useful, and almost always appropriate to use in explaining differences and similarities among various features on earth. The utility of the concept of region is further enhanced by its applicability to a wide variety of changing situations.

The Chernobyl accident provides an opportunity to illustrate the usefulness and flexibility of region as a geographical concept. Studies of the regions that include Chernobyl would differ greatly depending upon whether they were done before or after the 1986 nuclear explosion. Earlier studies may have concluded, after consulting an atlas, that Chernobyl was simply a small place in Ukraine—a republic in the Soviet Union. Chernobyl probably would not have been considered important enough to merit more than passing mention. Certainly, before 1986, nearly everyone with a reason to study Chernobyl would have assigned it little more than a brief note in a detailed description of Ukraine. Persons familiar with the locations of the world's nuclear power plants may have known Chernobyl for its nuclear facilities. Today, a study of Chernobyl would be vastly different from one that predated April 26, 1986.

To which region does Chernobyl now belong? Is the region of Chernobyl the devastated area, six to ten miles in radius, around the once powerful nuclear reactor? Is Chernobyl subsumed within Ukraine whose soils, now tainted, were known worldwide for their fertility and for providing grain to millions of Soviet citizens? Does the region of which Chernobyl is a part extend to Byelorussia, Latvia, Estonia, Lithuania, and other Soviet republics contaminated by the radioactive clouds from the disabled nuclear installation? Is the Chernobyl region now the entire earth? In truth, Chernobyl is part of all of these regions. The radioactive contaminants that spread from Chernobyl have touched all parts of the earth's surface. In this way, we and our home areas have become a part of the Chernobyl region.*

Acknowledgment
Alan Moore, geography major at Virginia Polytechnic Institute and State University, spent many hours preparing the maps in this chapter.

Teaching Activities

A major goal of this chapter is to foster thinking about the relationship between contemporary events and the five fundamental themes in geography (Committee on Geographic Education 1984). Contemporary events from around

---

*Coauthor R.W. Morrill, his wife Diane, and son Jeff were living in Finland from January 11 through June 17, 1986, during which the Chernobyl accident occurred. Chernobyl has a special meaning in their lives.

the world are reported in newspapers, magazines, and radio and television broadcasts. News reporting is a part of daily life. All events have geographic aspects and can be used to develop classroom activities that introduce and reinforce the fundamental themes in geography.

Activity One: *Global Crises*

Some events affect a large proportion of the earth's human population, influence large areas of the earth's surface, endure over time, and require the cooperation of many societies to solve or improve.

1. To begin the activity, divide the class into groups of 4 or 5 students.
2. Select a member of each group to serve as a recorder or reporter.
3. In group discussion, students identify and list all events and issues that may be considered global crises.
4. Students then refine and shorten the list by comparing each event or issue with respect to the following four criteria:
   a. A large portion of the earth's population is involved in the event.
   b. Large areas of the earth are influenced by the event.
   c. The event has effects that may endure for a long time.
   d. The cooperation of several countries is required to solve or improve the conditions related to the event.
5. Group recorders or reporters should report to the entire class the results of the small group discussions and compile a comprehensive list of global crises.
6. Students are asked to select one event that interests them and develop a paper or report that includes:
   a. *The location or locations of the event.* The event may have occurred at one or several locations. For example, a nuclear accident or major earthquake may occur at a single location, whereas a severe drought may affect many countries. Students should give latitude and longitude of the sites of the event (absolute location) and should describe what is near the sites of the event (relative location). Some events will be more difficult to locate than others. For example, warming of the earth's atmosphere may occur everywhere rather than at one or a few locations.
   b. *Places that are involved in the event.* In this instance, places are locations that have human settlements or have in some way been directly modified by human activity. For example, the eruption and explosion of Mt. St. Helens were natural environmental events that occurred at a specific and seemingly isolated location, yet many human settlements and activities were disrupted. Students should describe the human and physical characteristics of the places involved in the event.
   c. *Relationships within places or human-environmental interactions related to the event.* Human activities have intended and unintended consequences on natural environments. For example, cutting portions of tropical rain forests to obtain wood for building or clearing land for cattle grazing may have intended consequences of creating wealth for lumber companies and livestock owners but may also have unintended

consequences of destroying species of wildlife dependent on the rain forest. Students should become familiar with some of the complex positive and negative outcomes of human-environmental interactions.

d. *Movements related to the event.* The movements of people, goods, and ideas have produced interdependence and interconnectedness on a global scale. Nearly all peoples and places are influenced by events that occur in distant places. Diseases that originate in one area of the world threaten populations in many other areas. Unusual weather conditions may destroy entire crops in one country but contribute to bumper crops in another country. Such conditions may create serious food shortages in one place and surpluses in another. Movement of food supplies over long distances may be required to avoid starvation and malnutrition. Students should be able to identify examples of movements on the earth's surface and describe their importance to human well-being.

e. *Regions involved in an event.* Regions have unifying and distinctive characteristics. Defining a region involves identifying features that are distinctive of one area but may be totally absent elsewhere. For example, petroleum is abundant in some areas of the world but lacking in other areas. A petroleum-producing region may, through the actions of governments, restrict the flow of oil to countries dependent on outside sources of oil. Great hardships and even wars may result from such actions.

A concentration of wealth in one region of a country and poverty in another region may create serious political divisions within that country. Movement of people from the poor region to the wealthy region may cause turmoil. Characteristics of places that constitute regions are important to human life. Students should be able to identify and describe examples of physical and cultural regions and explain why regions form and change.

Activity Two: *Geography in the News*

Students may be asked to read a news article from a newspaper or news magazine or to watch a television program concerning an environmental problem. Examples of significant environmental problems may include (a) destruction of tropical rain forests, (b) depletion of the ozone layer in the earth's atmosphere, (c) persistent droughts in different areas of the earth, (d) pollution of the air, rivers, and oceans, (e) contamination of drinking water by widespread misuse of chemicals, (f) disposal of radioactive wastes from nuclear power plants and other uses of nuclear materials. Each of these problems may be described and analyzed using the five fundamental themes in geography. Students should be asked to apply the themes of *location, place, relationships within places, movement,* and *region* to one or more of the environmental problems. Written and oral reports and posters for display could be developed.

REFERENCES

"Chernobyl and Eastern Europe: One Year after the Accident." *Radio Free Europe: Research Background Report 67.* April 24, 1987.

Committee on Geographic Education, National Council for Geographic Education and Association of American Geographers. *Guidelines for Geographic Education: Elementary and Secondary Schools*. Washington, D.C., and Macomb, Ill.: Association of American Geographers and National Council for Geographic Education, 1984.

"Eastern Europe and Chernobyl: The Initial Response." *Radio Free Europe: Research Background Report 72.* May 23, 1986.

Edwards, Mike. "Ukraine" and "Chernobyl—One Year After." *National Geographic Magazine* 175 (May 1987): 595–631, 632–53.

Greenwald, John. "Deadly Meltdown." *Time,* May 12, 1986, 39–52.

Greenwald, John. "More Fallout from Chernobyl." *Time,* May 19, 1986, 44–46.

Ramberg, Bennett. "Learning from Chernobyl." *Foreign Affairs* 62 (Winter 1986–1987): 304–28.

Wise, John H. "Geographic Education and the Anticipation of World Events." *Journal of Geography* 79 (April/May 1980): 154–55.

CHAPTER 7

THE FUNDAMENTAL SKILLS OF GEOGRAPHY

George Vuicich, Joseph Stoltman, and
Richard G. Boehm

THE FUNDAMENTAL SKILLS OF GEOGRAPHY

As Dennis Spetz pointed out in chapter 5, one of the major problems for
school geography has been the small proportion of teachers with adequate
preparation in or understanding of the discipline. As a result, teachers often
consider exercises in map and globe skills *the* geography curriculum. Conse-
quently, they do not adequately address other fundamental skills in geography.
In addition, teachers frequently use maps and globes as learning props rather
than as laboratory tools, sources of data, and models for experimenting with
and testing ideas about the earth and its physical and human environments.

The skills of learning geography are designed to develop a sense of place and
space, one of the major goals of geographic education. A sense of place includes
knowledge about geographic locations as well as their physical and human
characteristics. It includes knowledge about the relationships that develop and
occur within and among places, the movements between and among places,
and the manner in which we use the concept of the region to organize and
understand the earth. In its broadest application, a sense of place and space is
an essential element one can apply across the social studies because all human
and physical events require space in order to operate.

The intellectual and social development of children depends upon a variety
of influences. Learning about place and space is similarly dependent. Life outside
school influences one's view of place and space. Children learn definite and
meaningful concepts of place and space through contacts with their peers,
adults, and various media. Similarly, work, travel, and life in general will
extend these concepts after formal school has ended. The development of a
sense of place and space in children in grades K–12 becomes the central component of the curriculum in geography.

Maps and globes are the fundamental tools for developing a sense of place
and space. The skills derived from the repeated and continuous use of maps
and globes can enrich all aspects of the social studies curriculum.

Professional Recognition of Geographic Skills

The professional literature in the social studies provides ample material that addresses the significance of geographic skills in the social studies curriculum. The National Council for the Social Studies devoted part of its 1953 and 1963 yearbooks to the theme of developing geographic skills in the social studies (Carpenter 1953, 1963). Each of the yearbooks devoted chapters to map and globe skills—"Interpreting Maps and Globes" in 1953 and "Developing a Sense of Place and Space" in 1963. Both yearbooks devoted attention to the other skills in social studies (charts, tables, graphics, reading and speaking, and locating and gathering information) that are also essential skills in geography. The 1959 yearbook, *New Viewpoints in Geography*, devoted one-half of the publication to geographic education skills and the application of concepts, generalizations, and skills to content-related issues (James 1959). The NCSS yearbook in 1965 focused on evaluation and included materials on how to assess geographic education skills in the social studies (Berg 1965). The most recent yearbook to address geography directly was *Focus on Geography: Key Concepts and Teaching Strategies* (Bacon 1970). It interspersed the skills of geographic education among chapters devoted to content topics. Although that yearbook touched upon the essential skills, they were not easily identified in comparison to the major content or applications elements of each chapter. In 1977, the NCSS Yearbook entitled *Developing Decision-Making Skills* included an important chapter focusing upon using maps, graphs, and direct observation in the decision-making process (Anderson and Winston 1977).

We can also document a second indication of the importance of geography skills in social studies by reviewing the first parts of many elementary and secondary school social studies textbooks. In the lessons devoted to skills development or the application of skills in processing the information and content of the text, most include map and globe skills development. Other publications deal exclusively with geographic skills and with learning sequences that focus upon maps and globes. These are generally available from major map companies and other educational publishers.

Numerous research studies deal with ways students develop abilities to use geographic skills. The most comprehensive reviews of research were those by Rice and Cobb (1978) and another much shorter review by Joyce (1987). However, there have been relatively few attempts to incorporate research results into a scope and sequence of geographic skills. The premier example of a research-based scope and sequence for map and globe skills was prepared by Winston (1984). Other scope and sequences for skills in geographic education have been suggested by Barton (1964) and the Task Force of the National Council for the Social Studies (1984).

Manson and Vuicich (1977) classified the skills associated with geographic education as demonstrating spatial competence in acting effectively and efficiently in geographic settings. Spatial competence, in their words, includes

such abilities as locating and orienting objects in space, adopting alternative perspectives of geographic phenomena, constructing appropriate models of portions of the earth's surface, and interpreting symbolic representations of places. For example, younger children can construct and use simple models and maps of their classroom, school, and community. Older children can map such distributions as population density and income. Younger children can learn to read simple maps of their school, neighborhood, and community, while older children can develop their map-interpretation skills by using road maps, atlas maps, topographic maps and, of course, the globe. Whatever the level of activity, four notions are central to spatial competence: location, direction, scale, and symbol.

SKILLS FOR HIGH SCHOOL GEOGRAPHY CURRICULA

The most recent statement regarding skills in the high school geography curricula is contained in *Guidelines for Geographic Education: Elementary and Secondary Schools* (Committee on Geographic Education 1984). Recognizing that skills scope and sequence statements traditionally addressed maps and globes, *Guidelines* emphasized that geographic education skills fall into five major categories:

- Asking geographic questions
- Acquiring geographic information
- Presenting geographic information
- Analyzing geographic information
- Developing and testing geographic generalizations

Guidelines goes on to provide detailed discussions of these skills. Each of the major categories includes map and globe skills but they are couched within the skills categories listed above. For example, maps are an important resource for acquiring geographic information. Similarly, maps are an important means of presenting geographic information. *Guidelines* also outlines ways in which maps may be used for acquiring information and the skills that students should develop in presenting information.

Asking Geographic Questions

Posing questions is one of the common denominators of classroom teaching procedures (*Harvard Education Letter* 1987) regardless of the teaching model employed—social interaction, information processing, behavior modification, or individual person (Joyce and Weil 1972). Furthermore, there are types of questions (or lines of questioning)—for example, convergent, divergent, informational, probing, leading, rhetorical, factual, recall, expository, inquiry, lower and higher cognitive level, to name a few—that are common to most, if not all, classroom teaching environments (Groisser 1964; Cornbleth 1977). Nevertheless, the framework or context in which questions are raised requires variations, some of which are important to the types of questions posed. One such context, of course, is a geographic one.

Take, for example, the following scene in a hypothetical social studies classroom. The day is Friday and the weekly discussion of international issues is on the Iran-Iraq war and the U.S. role in the Middle East. The goal of the teacher who is not particularly comfortable with geography is to "cover" the issue so that each student becomes generally familiar with the conflict. The line of questioning involves establishing at the outset a set of relevant facts and then moving to a higher *why* level. The teacher begins the lesson with what has taken place in the Iran-Iraq war and continues with questions of the following sort during the lesson:

- What do you think was the most important event in the Middle East during the past week?
- Who were some of the persons in the Middle East and what did they do that was newsworthy?
- What policy should the United States pursue in the Iran-Iraq war? Why or why not?

It is apparent that the teacher is employing a structured sequence of questions designed to take students through lower-level cognitive functioning into higher levels of reasoning employing a worthwhile topic. As such, many observers of such a classroom scene would probably nod approvingly as the lesson unfolds and conclude that the lesson was well taught.

Despite the orientation of the teacher involved, one would envision that he or she would at least use a world map (other than the one included in the news source) to locate the region and countries involved. In the scenario just described, it is quite conceivable that little or no geography would emerge except the cursory use of a wall map of the world.

Let's take the same issue with the same class but change the teacher to one interested in dealing with the topic in a geographic context. The sequence of questions might be somewhat as follows:

- What has taken place in the Iran-Iraq war?
- Who can come to the map and locate Iraq and Iran?
- What do you think was the most important event recently in the Middle East?
- Where did that event take place? Show us on the map, please.
- What policy should the United States pursue in the Iran-Iraq war?
- Does the geographic location of the United States relative to Iraq and Iran influence our policies toward those countries? How and why?

At first glance, there doesn't seem to be much difference between the two scenarios. However, the teacher who wants to emphasize the role of geography in the lesson establishes at the outset the spatial context of the topic by asking a straightforward *where* question. Thus the class reviews the event not only in the light of *who, what, when,* and *why,* but also *where.*

We can present a strong case in favor of including and emphasizing the elementary and basic notion of place and space *(where)* in most social studies classes. Reports that point out and lament the extent of geographic ignorance prevalent in all segments of our society abound and need not be reiterated here (*Dallas Times Herald* 1983; Kilpatrick 1987). It is important to reemphasize that knowing the location and the physical and human characteristics of places on the earth is only the beginning of geographical knowledge (see chapter 1). In addition, we need to mention and comment briefly upon the less discussed notion of dealing at higher intellectual levels with the geography of events and places. In the social studies classroom, we need more than ever to examine and study the significant geographical relationships that exist within and among places on earth.

Acquiring Geographic Information

Acquiring geographic information, like many processes, works best when students begin with simple procedures before moving on to more complex ones. The five fundamental themes of geography (Committee on Geographic Education 1984) provide guidance by suggesting that location and place should be addressed before the student can realistically understand human-environmental relations, movement or interaction, or regions. Thus, the first step in acquiring spatial information is to identify locations in order to create a framework or mental map upon which to add further geographical information.

Location information may be acquired in two forms—absolute and relative.

Absolute location refers to precise location and is normally determined on a map by some sort of grid. The most accurate grid is the world system of parallels of latitude and meridians of longitude that establishes for every place a specific location in degrees, minutes, and seconds. Other grids useful in identifying locations may not be as accurate—for example, road maps and city maps with alphanumeric grids. United States Geological Survey topographic maps, since they are part of the national grid, provide the best means for acquiring precise location information about local areas.

Relative location relates a place to some other geographic feature. Relative location draws heavily on our mental maps since we identify most places by their location in relation to other places. For example, San Marcos, Texas, is frequently described as being 30 miles south of Austin and 48 miles north of San Antonio. St. Louis is located in Missouri, near where the Missouri and Illinois rivers flow into the Mississippi.

By acquiring absolute and relative location information, we build basic mental maps that contribute to the understanding of higher-order geographic concepts. Location information answers the question *where,* which is basic to all geographic analysis.

Another fundamental way to gather geographic information is through careful observation of places and areas. Some observations are firsthand, whereas others are remote, relying on photographs or written or oral accounts. Several suggested techniques for observing places and areas follow:

1. *Using maps.* Maps show locations, movement patterns, and distributions. Such information suggests linkages and generalizations. Different types of maps show different things. Topographic maps show elevation, slope, drainage, and a variety of other physical features. The U.S. Geological Survey topographic sheets may be somewhat misnamed because they provide great cultural detail about small areas. Physical-political maps show physical features and political characteristics, such as state and national boundaries and the location of cities. Thematic maps show distributions, such as population distribution and density, agriculture, soils, vegetation, or economic development patterns.
2. *Field trips.* Carefully planned field trips can lead to a wealth of geographic information. Field trips are useful as early as the primary grades (see

chapter 8). It is often helpful to combine field trips with other useful information-gathering techniques, such as the photographic essay or the handwritten journal. Maps should be used to provide spatial order to this type of observation.

3. *Photographs.* When direct observation is not possible, geographic information may be gathered from air photos, satellite imagery, or even from snapshots. Such photos may be used along with direct observation to achieve a different perspective. Students in all grades are very much interested in seeing their immediate surroundings in an aerial photograph. Inevitably they find things they have missed in their routine travel patterns.

4. *Resource persons.* Resource persons may provide secondhand observation. For example, a class might visit the Director of Planning of the local municipality or invite that person to the classroom. Such persons bring insights into the spatial aspects of growth and development for the local community.

5. *Library.* A library also presents the opportunities for students to observe places through the eyes of others. Historical accounts, area studies, reference works, and documents all add bits and pieces of information to the geographic mosaic. Although library research is a valuable learning device, it should be used in connection with direct observation. Think of it as a source of supportive information that will help to round out one's perception of a place or area.

6. *Perception and values.* Observation can often be biased by the unique way each of us looks at the world. Poor persons may perceive things differently from rich persons. Hispanics may interpret information differently from Orientals. Each of us has a level of social consciousness and a set of values that make unbiased observation practically impossible. For example, one person may view a dam and a large lake as a recreational miracle, making boating, fishing, and water sports possible. Another may see the same features as a corruption of the natural environment, a heavy tax burden, and an instrument for the unnecessary displacement of hundreds of farm families formerly located in the river valley. The best method of observing places and areas is to use several of the suggested techniques to minimize distortion or bias. It may be useful in the classroom to conduct perception and values clarification exercises before observation actually begins.

Acquiring Geographic Data

In the intermediate and high school grades, we should encourage students to develop methods for acquiring data on places and areas. We should direct them to the sections of the library where published statistical information is available, such as census materials for the United States or world population information from the United Nations Demographic Yearbook. These are excellent sources for comparative data on birth rates, death rates, per capita gross national product, education level, and economic prosperity. The United States

Census materials can provide socioeconomic data for relatively small areas in your local community, such as census tracts.

Other ways of gathering geographic data use questionnaires or interviews. Students can develop simple questionnaires and pretest them on a group of students and adults before they administer them. Pretesting removes ambiguous questions and gives the developer ample time to rewrite appropriate questions. Try to ensure that no single group is overrepresented in the sample so as not to bias the results of the survey. All questions should be written so that respondents can answer simply. Yes or no answers are good, as are numerical answers. A popular form of questionnaire contains statements that use a Likert scale. For example, a questionnaire item might state, "This is a pleasant environment." Response choices might range from "strongly disagree" to "strongly agree" with a "neither disagree nor agree" in the middle. As a general rule, in interviewing, it is best to remove the kinds of questions that allow opinionated or open-ended answers. Precision in answers normally means more reliable information.

Presenting Geographic Information—Tables and Graphs

Geographers use tables and graphs almost as often as maps to present information and frequently use them in conjunction with maps to illuminate spatial phenomena.

The following hypothetical 5th grade geographical analysis problem demonstrates an appropriate use of tables and graphs in the classroom. The exercise begins with a general discussion to emphasize the location of residential areas in the students' home community or more specifically where people live.

Because the discussion will reveal that students need close observation of the problem, the teacher should plan to take the class on a field trip. The teacher should encourage students to observe carefully the residential areas and report their observations when they return to class. The students may notice that single-family houses farther away from the city center seem to be larger than those located near the center. A resource person from the local real estate agency can be invited to discuss the matter with the class. The resource person can give price information on 15 or 20 houses listed for sale. Students should then locate these houses on a city map and measure the distance of each of the houses from the city center. Results are shown in Table 1.

Interpreting Tabular and Graphic Information

Tables and graphs are important to geographic investigation because they allow students to organize data, develop tables and graphs, and examine the relationship between two variables. Often one of the variables will have spatial characteristics, such as distance or location.

To review the methodology, we can return to Table 1 and Figures 1 and 2. The table organizes information collected from at least two sources. We could draw various conclusions from a careful assessment of the data, but the logical step is to plot Table 1 data on a graph or graphs to show spatial relationships. We could have used a map but we would lose the clarity of the graphic relationships that lead students to the generalizations we seek. In examining

Table 1. SIZE AND COST OF SINGLE-FAMILY HOUSES IN LONGFIELD AND
DISTANCE FROM CITY CENTER

House Number	Size (Square Feet)	Price	Distance from City Center (Miles)
1	2170	$225,500	4.8
2	1450	79,800	6.5
3	1700	68,300	3.7
4	2250	98,700	4.5
5	1180	48,600	1.2
6	2674	128,100	5.0
7	1820	91,400	3.4
8	970	39,700	.8
9	1260	62,500	2.4
10	2700	137,900	5.5
11	1340	61,900	2.8
12	1150	89,100	7.0
13	1080	51,400	1.6
14	2222	117,500	5.2
15	1760	84,700	3.3

Figure 1, it is obvious that house size increases with distance from the city center. At least, this seems to be true up to about six miles out. Then house size decreases dramatically. The same pattern appears in Figure 2. House price seems to go up in direct relationship to distance from city center for about six miles and then there is a noticeable decline. However, the decline does not seem as precipitous as in the case of house size.

What can we learn from these spatial expressions of geographic data? Certainly the class's early observation that house size increases farther away from the city seems correct. But what happens at the six-mile limit? Is that the edge of town? Are the two houses beyond the six-mile mark unusual and unlike others in the same area? The graph on house price seems to reinforce some of these observations and to add some questions. Why does house price dip at the six-mile mark but not as severely as house size? Is this rural land? Are these two houses farmhouses with good-sized parcels of land attached? Might the houses be small but the land valuable for future development?

These questions and interpretations push the geographic analysis farther. New questions now arise and the class can seek more sophisticated answers.

The next exercise is a field expedition. Instead of using their general observations, students should prepare questionnaires to seek specific items of information. On the basis of Figures 1 and 2, students should be able to generate several hypotheses about the spatial distribution of houses in Longfield. Collection of further data and their careful analysis should lead to some interesting answers.

Figure 1
HOUSE SIZE RELATED TO DISTANCE FROM CITY CENTER, LONGFIELD

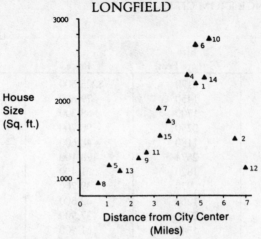

Table 1 is a good data source, but it is difficult to draw any conclusions. The following two graphs might help in elucidating the problem. One shows distance from city center related to house size, and the second displays distance and price. They are shown as Figures 1 and 2.

One could generalize the information in Figure 1 by fitting a line to the distribution of dots. Such a technique makes interpretation much easier. Figure 2 displays the information on price.

The information shown in Figures 1 and 2 leads to the process of geographical analysis. The graphs encourage speculation. They also provide convenient

Figure 2
HOUSE PRICE RELATED TO DISTANCE FROM CITY CENTER, LONGFIELD

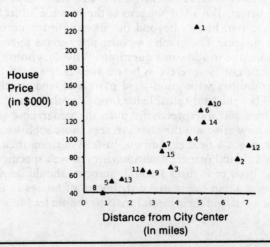

exercises for good writing. What do the graphs show? What can be interpreted from them?

Developing and Testing Generalizations: Local Geography

The tables and graphs in the Longfield example are tools that, like maps, can lead to generalizations or avenues for further investigation. We can test at least three generalizations.

Generalization One: *House size and price increase with distance up to six miles from city center.*

Generalization Two: *House size and value decrease beyond six miles.*

Generalization Three: *House size decreased more rapidly than house price beyond the six-mile mark.*

Testing the Generalizations

Generalization One: *House size and price increase with distance up to six miles from city center.*

Although this generalization seems to be true based on the information in Table 1 and Figures 1 and 2, fifteen observations are too few to justify a final conclusion. Also, at no time did we map the location of the 15 houses to see whether they clustered in one area.

The following procedures are suggested:

1. Divide the city into quadrants.
2. Ask a real estate agent to provide size, cost, and location information on 15 houses in each quadrant.
3. Graph the information following procedures similar to those used with Figures 1 and 2 (house size related to distance).
4. Look for significant differences among the eight graphs and between each of the graphs and the original Figures 1 and 2.

If all the graphs are similar, we can tentatively accept Generalization One. If they are not similar, we might seek other generalizations. For example, suppose the graphs for one quadrant showed no relationship between house size, cost, and distance. The class might conclude that Generalization One was accurate except for one quadrant of the city.

Generalization Two: *House size and value decrease beyond six miles.* This suggests that this area is near the outer edge of the city where rural areas begin.

The class can test this generalization either by a field trip or by careful analysis of a city map. It is sometimes useful to obtain a map showing the city limits and the city's extraterritorial jurisdiction. It would also be useful to observe some decrease in house size and value on graphs for all four quadrants. Once all this information becomes available, the class, after discussion, could accept or reject Generalization Two.

Generalization Three: *House size decreased more rapidly than house price beyond the six-mile mark.* This suggests that the higher value may be related to land that has potential for development into extensions of the suburbs.

This generalization is complex and clearly needs further investigation since the decrease in value and size of houses was not noticed by the students in their initial observations. The following procedures are suggested:

1. Investigate the eight graphs to see whether the pattern is the same as in Figures 1 and 2.
2. Obtain a zoning map to determine whether land along the rural-urban fringe has been rezoned for commercial or residential development.
3. Develop a questionnaire asking people in the six- to seven-mile zone about future use of their land.
4. Develop a second questionnaire asking several real estate agents about the future of this land.
5. Invite the Longfield Director of Planning to come to class to discuss what you have found in the data collection, display, and interpretation phase of your geographic analysis. Ask the director about any new real estate or commercial developments in the six- to seven-mile zone.

Based on this additional information, the class should be able to accept or reject Generalization Three. Of course, new information leads appropriately to new questions and possibly new generalizations.

Presenting Geographic Information—Maps, Globes, and Animated Displays

Maps, globes and animated displays can present a wealth of geographic information. Maps offer a great opportunity for presenting geographic information because they enable students to visualize the spatial characteristics of the information presented. In the vocabulary of geographers, 'spatial' is an important term, but few other people use or recognize its meaning. 'Spatial' means "in relationship to the surface of the earth, or pertaining to space." 'Space', in a geographic sense, generally refers to space directly on the earth's surface, but may also refer to the three-dimensional surface of a city, a shopping center, or similar places that may be mapped. If we extend the discussion of 'spatial', we discover that it refers to one method for classifying information about the earth. Another way to classify information is the temporal or historical, i.e., through time. Maps and globes are important sources of information when considering spatial or geographic questions about the earth. Maps and globes also portray the intersections of the spatial and temporal dimensions of the human experience. For example, a map showing the North American westward migration of people also portrays how the spatial distribution of the population has changed over a period of time.

Information on the physical and human characteristics of places presented on maps can have important implications for its users. For example, people who travel from one place to another in the world can benefit from knowing the location or spatial distribution of highly infectious diseases. Malaria affects large numbers of people in certain parts of the world, but people in other places are free from the disease. The map (Figure 3) shows the spatial distribution of malaria. By referring to this map, a business person, tourist, Peace Corps worker, exchange student, or airline flight crew would be able to determine quickly whether contracting malaria is possible in the region or place that is their destination. The map presents general information about places where malaria is a problem, and a traveler to those places should determine how to avoid contracting the disease. By examining even larger and more detailed maps of specific countries, a person can pinpoint accurately the particular disease vectors in that country.

The map of the spatial distribution of malaria also includes a time dimension by depicting regions and places where malaria has disappeared, where it has been eradicated, or where it has never occurred. The Central American region is an example of where the spatial information and temporal information on the map converge. Of the countries that share the region, malaria transmission occurs in each of them, but the disease is not evenly distributed. Why has malaria become a health problem with limited risks in Costa Rica and southern Panama, whereas most of the region continues to have malaria transmission among its population? Does it seem feasible that malaria was initially a problem in every country except those two? A review of a physical geography map shows that Central America has a variety of landform features but this variety repeats itself in each country. This suggests that the mosquito, the carrier of malaria, is at home throughout the region. The present spatial distribution suggests, therefore, that at some time in the past there was a successful plan to control the spread of malaria in Costa Rica and Panama and that such a program either was not started or was not successful in the rest of the region. That question is open for further inquiry. For travelers, the information on the map requires them to take special steps to avoid malaria while visiting Central America.

Written and Oral Presentation of Geographic Information

A variety of geographic information can provide data that can develop and reinforce students' written and verbal skills. When students present geographic information, they should be required to include graphic representations such as maps or aerial photos in their reports. Of the two, the map will probably be easier to include. Very young students could sketch maps from memory. At an early age, students should be made aware of how to use maps, both as ways to describe how a set of data is distributed (for example, houses, buildings in a neighborhood, play equipment on a playground, desks in a classroom) and to determine both the existence and degree of relationships that exist between sets of objects.

For older students, the reports could include commercially produced 8½-by-11-inch base maps on which they can map relevant data. They can present these distributions in a number of ways—for example, by simple coloring techniques or by a variety of gray tones to show variations in the distribution. When students present oral reports, they should use the maps as visual aids during the presentation by making and using overhead transparencies of them.

Figure 3
DISTRIBUTION OF MALARIA

MALARIA: STILL A PROBLEM IN MUCH OF THE WORLD

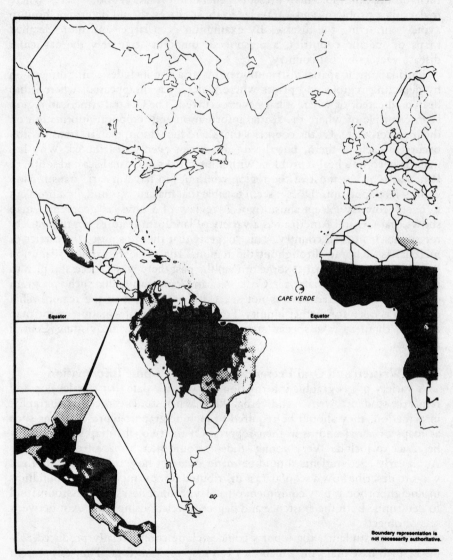

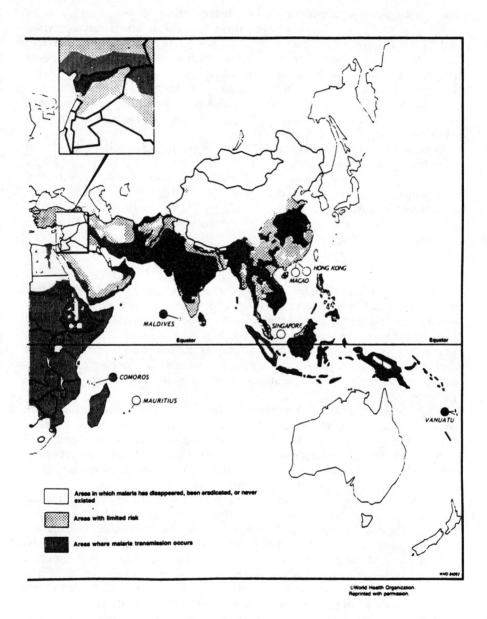

Areas in which malaria has disappeared, been eradicated, or never existed

Areas with limited risk

Areas where malaria transmission occurs

©World Health Organization.
Reprinted with permission.

Patterns of the distributions—which are important features in these maps—should have sufficient contrast to be clearly discernible.

Analyzing Geographic Information: Interpreting Maps

All map users should learn early that it is impossible to take a curved surface and flatten it without introducing some change (error) in the original surface. In mapping, this means that whatever is portrayed on a two-dimensional surface is distorted in some fashion. The size of the area mapped is also a factor. The smaller the area, the less the distortion—of any type. In creating a two-dimensional map, cartographers speak of distortion occurring to one or more of these critical elements of a map: angles, areas, distances, and direction (Robinson et al. 1984; Robinson 1987).

In grappling with the problem of distortion, cartographers have developed complex mathematical ways of correcting these errors. One category of map projections corrects for or minimizes distance or scale error. Another group of map projections corrects for direction error, another for areas, and still another for shapes. Because there is no conceivable way to deal with all the distortions simultaneously, the map user must exercise care in selecting the projection most suitable for each mapping project.

Recognizing Patterns

Geographic information includes any data represented on maps. Such information is shown in two general ways—as points or as patterns. Points show sets of data having a small areal extent, such as cities. Patterns show the location of a set of data covering a wide geographic area, such as climatic characteristics. Points and patterns are related because the regular occurrence of similar point data at a large number of locations results in a pattern. The incidence of malaria at a large number of specific locations or points on the map results in a global pattern of the disease.

Social studies instruction has generally focused its geographical information on point data. The locations of cities, rivers, and boundaries are examples of this type of spatial information. Pattern information, on the other hand, provides a basis for the identification of geographic regions. In its broadest sense, the region is a way to classify spatial information in order to conceptualize complex reality in meaningful and manageable ways.

The map showing the assessment of the status of malaria as a communicable disease is a way of showing the connection between point and pattern information. The data used to compile this map were the locations (points) of malaria cases reported to the World Health Organization. Points on this map are best considered as places, such as Tokyo, Vanuatu, Nairobi, and so on. At its most general level, the map provides information regarding reports of malaria at specific locations around the world. However, because it is difficult to keep track mentally of all the specific locations, we need to classify those locations spatially to comprehend them. The pattern on the map of malaria holds greater meaning for students than describing it verbally.

Determining Relationships

The incidence of malaria is closely related to several different environmental factors, such as the presence of surface water, the necessary daily and annual temperature ranges, and the habitat of the anopheles mosquito that is the host for the parasite that causes malaria. The spatial patterns of these environmental factors show a close and readily discernible relationship with the pattern of malarial incidence. The pattern on one map must be compared to the pattern on other maps to help determine whether the two sets of information appear to be related.

The pattern on the map of malaria transmission in Egypt is concentrated in a distinctive way. The pattern may be called the malaria region within Egypt. When we see such a distinct pattern on the map, then we can ask other types of questions:

1. What are the human characteristics of this region? Answer: Nearly all of Egypt's population lives in this region, and the population density is high. This information comes from a population distribution and density map.

2. What are the physical characteristics of this region? Answer: The Nile River flows through this region. Much of the region was once subject to annual flooding, but flood waters are now controlled by a series of dams, the largest being the Aswan Dam. This information comes from historical accounts of the Nile River and recent accounts of the water control projects. Detailed maps of Egypt will also show the locations of these projects.

3. How do people use the land in the Nile Valley? Answer: Much of the agricultural land in this region is irrigated by water from the Nile. There are fields with irrigated crops, large numbers of irrigation canals, and drainage canals. Many pools of stagnant water dot the surface of the region and provide ideal breeding sites for the mosquito. This information comes from a number of different sources: land-use maps of the Nile River valley, written information about the problems with mosquitoes, and pictures or aerial photographs showing irrigation canals, ponds with water, and flooded fields.

Examples abound for comparing relationships presented by different maps. A classic example is the relationship between precipitation and vegetation in many parts of the world. The relationship between precipitation and vegetation may also be extended to land uses for agriculture. Maps contain many patterns that we can explain in part by examining other maps and looking for relationships between different kinds of information to help explain why locations, places, and regions have certain characteristics. These move students beyond the first level of map reading toward explanation regarding the patterns they observe on maps.*

---

*Similarities between or among patterns on maps are not sufficient to determine whether or not a *functional* relationship exists between or among the phenomena studied. Here is where additional information or reasoning becomes necessary. To use our earlier examples, one needs to reason logically or obtain information from other sources to discover the links between people and malaria or between vegetation and climate. In some cases, these linkages are fairly obvious; in others they become quite complex.

Developing and Testing Geographic Generalizations:
Global Geography

Generalizations in geography at the global scale are usually related to the earth's systems. The earth's physical and ecological systems are interconnected to other components of the whole and are governed by physical laws. Some of the larger elements of earth systems are latitude and its relationship to climate and vegetation, oceanic circulation patterns and their resulting effects on weather, and the relationship between elevation and air temperature. Other systems, such as economic and social systems, may also be used to develop generalizations, but they require different sets of assumptions as a result of the human variables they consider. The procedure that students follow is described in the following example.

Developing generalizations regarding the global distribution of malaria. Begin developing geographic generalizations with maps. One of the first and most useful of the global systems includes latitude north and south of the Equator with its effects upon seasons and climates.

The following generalization regarding malaria may be developed by examining the map and studying the patterns of occurrence and asking the following two questions. First, where and what are the regions with the highest incidence of malaria? Second, where and what are the regions that do not have any reported cases of malaria? Those questions lead students to recognize that the presence and absence of malaria in a certain region is because of . . . . A generalization is the next step.

Generalization One: Malaria is related to the physical characteristics of places.

Testing generalizations regarding the global distribution of malaria. We made the foregoing generalization by using information from one map. Students should test the generalization by referring to several different maps usually available in a classroom atlas or text. Testing the generalization first requires comparing the conditions of the physical environment at places or in regions with malaria.

1. One outstanding pattern for the distribution of malaria in the map is the presence of the disease in latitudes near the Equator and its absence in the higher latitudes. Malaria is a problem both north and south of the Equator, but ceases to be a problem at approximately the same latitude in both the northern and southern hemisphere.

2. Large malarial regions on the continents of Africa and South America are also regions of high precipitation. This may be verified by looking at the map and comparing the distribution of malaria to rainfall on a precipitation map.

3. One can observe that people in some regions with less rainfall also have malaria, e.g., the Arabian Gulf and the west coast of Mexico. It is necessary to examine maps of the physical characteristics of those regions to determine whether there are sources of water in which the mosquito can breed, e.g., seasonal rains and irrigation in oases communities.

4. There is also a distinctive pattern of malaria along the western part of South America and in Mexico and Central America. Malaria does not occur for the most part in the Andes Mountains region. What is the pattern in other

major mountain regions? The Himalayas and the Plateau of Tibet are free of malaria. To obtain answers to this question, it is necessary to know the characteristics of the anopheles mosquito and the effects of elevation on its survival.

It is possible for students to expand the development of generalizations using the map. Additional generalizations may focus on the human characteristics of places, including economic development. Several additional generalizations might appear as follows:

1. People in places with *developed* economic systems face few or no risks from malaria.

2. People living in places with *developing* economic systems often face risks from malaria because of varying stages of economic development within a region.

We can test these generalizations by obtaining additional information from an atlas or data table presenting economic development. Then we can accept, reject, or modify the generalization. The exercises demonstrate how the student engaged in developing and testing generalizations in geography goes beyond the basic skill of map reading by using the map as a tool or resource in searching out the solution to a problem or an issue.

SOURCES FOR GEOGRAPHIC INFORMATION

Standard Sources

Geographic information includes both qualitative and quantitative data about location. We have emphasized quantitative data since these lend themselves to description and analysis through mapping. Many excellent sources for geographic information are readily available for use. They include a variety of international, national, regional, state, and local sources.

The United Nations is an unmatched source of statistical data at the international level. Specific information regarding UN documents is available through *UNDOC: Current Index, United Nations Documents,* which is issued four times annually. For subscription information regarding this publication, write to: United Nations Publications, Room DC2-0835, New York, N.Y. 10017. Also excellent sources for data on foreign countries are the foreign consular offices in the United States. Addresses for these are available from the Superintendent of Documents, U.S. Government Printing Office (U.S. GPO), Washington, D.C. 20402 (phone (202) 783-3238), in the publication entitled *Foreign Consular Offices in the U.S.* Less well known but equally helpful is *Key Officers of Foreign Service Posts, Guide for Business Representatives,* Department of State Publication 7877, and also available through the U.S. GPO.

For national data, the annual *Census Catalog and Guide,* U.S. Department of Commerce, Bureau of the Census, for sale by the Superintendent of Documents, U.S. Government Printing Office, Washington, D.C. 20402-9325, is a publication all school libraries should have. Also useful and available through the Superintendent of Documents on a quarterly basis is *Government Periodicals and Subscription Services.*

Most states have a wealth of useful information from their respective Departments of Natural Resources or Geological Survey Divisions. Similarly, state library services regularly publish catalogs of available documents.

Electronic Data Sources

Electronic data sources are those that may be used with computers. In recent years, opportunities have increased greatly for teachers to obtain information in electronic form. Teachers have several alternatives for using computer data sources in working with geographic information.

First, there are several data base files that may be used for developing one's own data base. This normally entails selecting the information that teachers or students deem essential and entering it in the data file in a standard format. PFS File is one example of such a file.

Second, there are data bases that are distributed by software companies. It is advantageous to obtain a data base that may be updated or have information fields added as necessary. Three data bases that have a high degree of utility for geographic information are produced by Active Learning Systems, 5365 Avenida Encinas, Suite J, Carlsbad, California 92008. The first is "One World Countries Database." It contains 33 sets of information about 178 countries. The second is "USA Profile: Social and Geographical Database"; it contains 35 sets of information about each of the 50 states. The third database, "Hometown: A Local Area Study," is a worksheet on which data about the home town or local area may be arranged and developed into a data base. In each of the three data bases, the computer program enables students to compare and analyze information easily and quickly. Other commercial software data bases are also available and are advertised in computer journals and lists of educational software.

The United States Bureau of the Census provided certain data from the 1980 census on diskettes for use in microcomputers. Many of those data are similar to the printed volumes of census information but have the advantage of being easily accessible to students through a microcomputer. By using those data diskettes, specific data regarding population may be located and printed. These are available from the *Current Census Catalog and Guide.*

A comprehensive source of information regarding the availability of electronic databases is the *EPIE Journal* (*T.E.S.S. The Educational Software Selector,* 1986). Computer periodicals written especially for teachers often carry reviews of software, including data bases. The Clearinghouse for Social Studies/Social Science Education at Indiana University (2805 East Tenth Street, Suite 120, Indiana University, Bloomington, Indiana 47405) publishes materials on a regular basis, some of which make reference to using computers and data bases in teaching geography.

REFERENCES

Anderson, C., and B. Winston. "Acquiring Information by Asking Questions, Using Maps and Graphs, and Making Direct Observations." In *Developing Decision-Making Skills,* Forty-Seventh Yearbook, edited by D. Kurfman. Washington, D.C.: National Council for the Social Studies, 1977.

Bacon, P., ed. *Focus on Geography: Key Concepts and Teaching Strategies,* Fortieth Yearbook. Washington, D.C.: National Council for the Social Studies, 1970.

Berg, H., ed. *Evaluation in Social Studies,* Thirty-Fifth Yearbook. Washington, D.C.: National Council for the Social Studies, 1965: 100–14.

Barton, T. "Geography Skills and Techniques." In *Curriculum Guide for Geographic Education,* edited by W. Hill. Macomb, Ill.: National Council for Geographic Education, 1964: 51–71.

Clearinghouse for Social Studies/Social Science Education, 2805 East Tenth Street, Suite 120, Indiana University, Bloomington, Indiana 47405.

Carpenter, H., ed. *Skills in Social Studies,* Twenty-Fourth Yearbook. Washington, D.C.: National Council for the Social Studies, 1953.

Carpenter, H., ed. *Skill Development in Social Studies,* Thirty-Third Yearbook. Washington, D.C.: National Council for the Social Studies, 1963.

Committee on Geographic Education, Association of American Geographers and the National Council for Geographic Education. *Guidelines for Geographic Education: Elementary and Secondary Schools.* Washington, D.C., and Macomb, Ill.: Association of American Geographers and National Council for Geographic Education, 1984.

Cornbleth, Catherine. "Using Questions in Social Studies," How To Do It Series. Series 2, No. 4. Washington, D.C.: National Council for the Social Studies, 1977.

Dallas Times Herald. "American Education: The ABCs of Failure." Dallas: Times Herald, 1983.

Groisser, Philip. *How to Use the Fine Art of Questioning.* Teachers Practical Press, 1964.

Harvard Education Letter 3, no. 3, May 1987.

James, P., ed. *New Viewpoints in Geography,* Twenty-Ninth Yearbook. Washington, D.C.: National Council for the Social Studies, 1959: 112–210.

Joyce, W. "Research Tells Us That. . . ." *Michigan Social Studies Journal* (February 1987): 77–80.

Joyce, Bruce, and Marsha Weil. *Models of Teaching.* Englewood Cliffs, N.J.: Prentice-Hall, 1972.

Kennamer, L. "Developing a Sense of Place and Space." In *Skill Development in Social Studies,* Thirty-Third Yearbook, edited by H. Carpenter. Washington, D.C.: National Council for the Social Studies, 1963: 148–70.

Kilpatrick, James. "American Students' Knowledge of Geography Is Abysmal, Embarrassing." *Kalamazoo Gazette,* May 31, 1987.

Kohn, C., et al. "Interpreting Maps and Globes." In *Skills in Social Studies,* Twenty-Fourth Yearbook, edited by H. Carpenter. Washington, D.C.: National Council for the Social Studies, 1953: 146–77.

Manson, Gary, and George Vuicich. *Toward Geographic Literacy in the Elementary School.* Boulder: Social Science Education Consortium, Inc., 1977.

McCune, G., and N. Pearson. "Interpreting Material Presented in Graphic Form." In *Skill Development in Social Studies,* Thirty-Third Yearbook, edited by H. Carpenter. Washington, D.C.: National Council for the Social Studies, 1963: 202–29.

Rice, M., and R. Cobb. *What Can Children Learn in Geography? A Review of Research.* Boulder, Colorado: Social Science Education Consortium, 1978.

Robinson, Arthur. "Reflections on the Gall-Peters Projection." *Social Education* 51 (April/May 1987): 260–65.

Robinson, Arthur, et al. *Elements of Cartography,* 5th ed. New York: John Wiley & Sons, 1984: 83–84.

Task Force of the National Council for the Social Studies. "Skills in the Social Studies Curriculum." *Social Education* 48 (April 1984): 259–61.

T.E.S.S.: The Educational Software Selector. New York: Teachers College, Columbia University, 1986.

Winston, B. *Map and Globe Skills: K–8 Teaching Guide.* Macomb, Ill.: National Council for Geographic Education, 1984.

CHAPTER 8

GETTING GEOGRAPHY INTO THE CURRICULUM

PART 1: PURSUING A DECALOGUE

James F. Marran

All geographers know that everything is somewhere. Place and location are as much a part of life's inescapable realities as are the physical and cultural dimensions of spatial relationships. Students of geography are always trying to fathom the variety and complexity of their environments through observation and analysis. *Where* and *why* questions are as essential to the geographer as the hand calculator is to the mathematician and the time line to the historian. Therefore, the geographer's special angle of vision is appropriate to any setting, no matter how mundane or commonplace. Consider the following scenario.

Waiting in the bitter cold invites seeking all manner of distractions to lessen the discomfort. Upper middle westerners are especially skilled at finding such diversions. On a recent near-zero morning in mid-March when the world should have been on the threshold of spring but was still trapped in an arctic freeze, I stood huddled with a dozen or more other winter-numbed commuters, under a heat lamp on the Belmont Avenue elevated platform of Chicago's North Side waiting for the train to Howard Street. As is almost always the case on such days, trains were running in clusters at about 20-minute intervals. Tired of scanning the billboards advertising Caribbean vacations and of the satisfaction that only a cigarette can bring, I turned to reading the graffiti.

Amidst the scrawls where Manny, Tom, or Mel had declared undying affection for Margie, Sue, or Holly and the Latin Kings had proclaimed their ubiquity in all of Chicago's mid-north neighborhoods, there was one uncommonly curious squiggle over a Citicorp ad, printed in neatly formed large, uppercase letters. "To be or not to be ain't much of a choice," it proclaimed.

Immediately, I began to warm to the message. What wisdom! I thought, and how applicable to so many of life's situations! Although graffiti may be humorous, insightful, and timely, they are seldom profound and almost never philosophical. In a paradoxical world of increasing informality but declining civility, wall talk and bumper-sticker messages have become devices urban folk develop to communicate with each other without ever having to become involved. Even among the ruins of Pompeii, graffiti were the big town substitute for down-home small talk. But the Shakespearean parody I read that morning was beyond sophistry and cleverness because of what it suggested to my schoolmaster's concern for effecting changes in the curriculum. In fact, that was where my mind was at the time. How could I encourage more geography instruction across the curriculum in a way teachers would respond to with something other than a few token lessons on map skills? Certainly, to include geography from a sense of wilting duty rather than solid commitment is not much of a choice. Like psalmists, graffitists are often prophetic!

Seasoned practitioners know well what every rookie learns early in the game: that changing what teachers teach is about as easy as amending the federal Constitution. Indeed, amending the Constitution might be easier because the reluctance of teachers to change is rooted in a tradition that predates the Constitution by a millennium or more. There is an Aristotelian constancy to what happens in classrooms because teaching is such an insular and solitary activity. The instructional process is so personal and fraught with so many priorities that the suggestion of change in content comes more as a threat than a challenge. To suggest, for example, that geography ought to be more evident across the social studies curriculum may meet with nods of agreement at staff meetings, but specific suggestions for implementation are likely to receive one of two reactions. People will respond either with indifferent and disdainful stares or with loud protests that there is already too much to "cover." The time to begin alterations in the curriculum is certainly not *now* but at some time in the vague *later,* which is always safe because everyone knows that *later* is *never.*

But modifications do take place in the schools. Indeed they must. When they come, changes may be wrenching and as thunderous as the footsteps of Jack-and-the-Beanstalk's giant. In spite of all the anxiety involved, change can be effected if its agents recognize that the cardinal rule in the school setting is not to do anything unilaterally. Everyone involved must be committed to both the concept and the process. Teachers cannot expect success if they blindly buck administrative hierarchies in their districts, nor can boards of education or administrators expect success if they decree change through bureaucratic ukases. Manipulating change through devious and clandestine means will not work either. Enduring modifications of the school curriculum can occur only when those who want change undertake it methodically and understand that there are principles that, like taxonomies, spiral upward to that point where all the parts become the whole. To get a geography program under way or to renew one that is flagging, those involved must simultaneously pursue the following decalogue as if it were written in stone. To do so will insure that there will be more options than "to be or not to be."

1. *Know the system.* Every district has a table of organization designed to ensure the orderly operation of its schools. This includes not only the responsible expenditure of public funds but the regular assessment of the curriculum through some formal system of evaluation of the teaching and learning process. Those who want to encourage change in content or methodology must know how the system works. That means identifying the decision makers and becoming familiar with their styles. Any strategy for change must operate within the design structure of the school district. Otherwise, it is predestined to fail. No

matter how well-meaning or important, proposals for change that circumvent or ignore how the district functions will be considered either amateurish or anarchical or both.

2. *Have a plan.* Good ideas often founder because they are never fully developed. Many sound innovations die at the faculty room coffee table because nobody takes the time to put them in writing. Every proposal must be made formal. That means including a statement of objectives, a description, explanations about implementation, a method of evaluation, and cost analysis (if that is applicable). Most districts have calendars for submitting course proposals or presenting grant applications for curriculum development. Meeting such deadlines is essential. Failure to do so means being barred on a technicality. Usually, there are specific forms districts have developed to standardize and expedite the process.

In short, using the system effectively through planning is as important as knowing its structure. Successful change almost always comes when teachers and administrators operate in a climate of cooperation and goodwill. That occurs when a specific plan is in hand for reference and discussion.

3. *Locate a resource base.* Even the best laid plans sometimes falter because the initiators fail to establish a support base of consultants and advisers. Good plans for curriculum change become improved when they are submitted to an objective review process before they go to the decision makers. But with limited budgets (or none at all in most instances) for such purposes, where can teachers turn for help?

Surprisingly, there are many competent people available simply for the asking. They can be found in the geography departments at local colleges and universities, in nearby community colleges, through local and state geographical societies, and through the nationwide network of consultants maintained by the Geographic Education National Implementation Project (GENIP) and the several state alliances of teachers under development by the National Geographic Society (NGS).[1]

Such references give not only added legitimacy to a proposal but insure expert advice based soundly on the scholarship of the discipline as well as on the most current educational research. With such reinforcement, decision makers are likely to find it difficult to deny carefully worked out and soundly endorsed initiatives.

4. *Include a system of evaluation.* There are two nagging questions surrounding every curriculum proposal: (1) Will it work? (2) If it does work, how will we know? Accountability is an inseparable component of every plan for change and so an evaluation procedure must be included. It need not be complex. In fact, the simpler it is, the more appealing it is to all involved. But it is important to include methods for the consistent measurement of the quality of the proposal. This might be anecdotal (on the part of both the students and the teacher) or include more formal systems of evaluation (e.g., tests, checklists,

---

[1] Information about networks in geography can be obtained through GENIP, 1710 Sixteenth Street N.W., Washington, D.C. 20009. Materials on the state alliance programs can be obtained from The National Geographic Society, Washington, D.C. 20036.

questionnaires. An effective curriculum must, above all else, be teachable—it must be designed to work for every teacher and not just for the developer. Positive results accruing from objective evaluation will insure that such is the case.

5. *Identify appropriate materials.* Sound curricular planning must be documented by academically credible research data. Otherwise, proposals, no matter how carefully developed, can easily be challenged by critics and dismissed quickly as shallow. Local geographers and education specialists can be helpful particularly in recommending references and in directing planners to data searches as a guarantee that the proposal will have the strongest possible research base. Being able to show how current research literature supports a proposal is an effective endorsement. Such annotations signal that the developers have done their homework and expect to be taken seriously.

6. *Plan in-service opportunities.* Unless teachers are convinced that a curricular change makes sense and complements their work in classrooms, failure is certain. To forestall such rejection, every proposal of value must include an in-service component. These sessions need to be well planned, to the point, and clearly presented. The most effective in-service sessions are those that involve teachers as active participants rather than as passive listeners. Converts to change can be won over if they are convinced that what is proposed is better than what it replaces.

7. *Keep the public informed.* Too often, teachers forget that the schools belong to the people who support them. Keeping the public apprised of curriculum developments in their schools encourages both goodwill and program endorsement. Regular press releases, newsletters, forums, panels, and workshops are more than public relations efforts. They are informational vehicles that cultivate pride in the local schools, one of a community's richest assets. Accurate information about school curricula makes good press and builds solid school-community relations. It is better to see a headline in the local paper that reads "Schools Plan to Beef Up Geography Programs" than one that proclaims "Critics Assail Schools for Ignorance about Geography."

8. *Be willing to compromise.* Like politics, curricular change involves the art of the possible. Success may not come as a full loaf, but being willing to make concessions to get a plan implemented is better than making no progress at all. Being intransigent about a proposal can easily spell its defeat. Giving a little by meeting critics halfway is not equivalent to selling out. Because schools are institutions of democracy, democratic processes must prevail. In the long run, pragmatism, not purism, shapes the program of studies in the schools.

9. *Make simplicity a priority.* Complexity can paralyze change in the schools. Proposals composed in jargon and rampant with educationese can bewilder even the most ardent supporter. Ideas that are expressed simply are far more convincing than those contorted with linguistic complexities. How much better, for example, to state that "the purpose of this proposal is to improve student map-reading skills in the middle school" than to write that "it must be assumed that to impact positively on student development in the awareness of spatial relationships, middle schoolers must become more adept in the reading, analysis, and interpretation of maps."

10. *Be prepared to start over.* As is the case in almost any endeavor, timing is important in curriculum change. Sound proposals sometimes fail to make the grade because their time has not yet come. Their strengths, however, lie in their abilities to serve as precursors for similar proposals that may emerge sometime in the future. Proponents of global studies more than a decade ago are the best examples of how persistence and patience pay off in the school setting. In the 1960s and early 1970s, despite their best efforts, there was only marginal interest in the global education movement. Now, although still controversial, the fruits of their efforts are evident in school curricula throughout the country. Courses across the social studies spectrum have been internationalized, and the importance and reality of global interdependence are evident in textbooks, teacher guides, audiovisual materials, and student/teacher resource materials. Some good ideas need a gestation period. Their proponents must therefore exercise prudent tenacity and operate as responsible persuaders for the curriculum change they represent.

As the graffitist at the Belmont Avenue station noted, polarization does not provide much of an option. Having geography or not having geography as a school subject is clearly not the issue. Ensuring that legitimate geography is represented in the K–12 social studies curriculum is essential. Students in history and the social science courses will have a limited perspective if they are denied awareness of the principles of geography. Solid and realistic curriculum planning will not only broaden options but improve geographic literacy as well.

PART 2: IMPLEMENTING A GEOGRAPHY PROGRAM

Salvatore J. Natoli

Previous chapters in this bulletin provide sufficient rationale for implementing a geography program in schools. This section will help outline some specific attributes of geography that will require special attention in implementing your objectives.

Of all the general liberal arts subjects taught in the schools, geography depends heavily upon visual teaching aids ranging from maps, all kinds of graphics, motion pictures, videocassettes, videodiscs, slides, and filmstrips. Most classrooms today have equipment for using films, filmstrips, videocassettes, and transparencies. There is a large body of literature that deals with effective use of these audiovisual aids. In addition, microcomputers are becoming increasingly common in classrooms. Software for computer-assisted instruction in geography is available from a variety of sources (some of which are listed at the end of this chapter), although its quality and utility for geographic learning varies considerably. The *Journal of Geography*, published by the National Council for Geographic Education (NCGE), and *Social Education*, published by the National Council for the Social Studies (NCSS), regularly publish reviews of software and these can be helpful for the potential user before any expenditure is made.

Of all the visual aids required for effective geography instruction, the map is without peer as a necessary learning and research tool for teaching and learning geography. The map, however, is an abstraction, a compilation of real world data onto a plane, or in the case of a globe, onto a spherical surface. Perhaps the only rationalization for requiring maps to portray and explain geographical phenomena and data is that geography deals with the real world. Thus, the greatest challenge to the teacher of geography is to plan for and assemble the necessary map and audiovisual tools that can translate effectively the various levels of the cone of experience—words, sound, sight, and simulated experiences—into experiences in the real world. This is the beginning of geographic education (Eliot-Hurst 1973).

Maps are essential for geographic understanding. Earth locations occur in two-dimensional space, so the symbols of these locations must be represented in two dimensions. In the sequence of geographic learning, maps are indispensable for collecting data by earth locations, analyzing areal and regional information, and formulating generalizations about spatial relationships. One cannot teach or learn geography without using a map or even several maps simultaneously in order to view spatial relationships among several different mapped distributions. Thus, desk and wall maps, globes, and atlases create distinctive requirements for classroom use and storage.

Some geography activities resemble closely other academic programs that use visual displays such as art or graphics. Others require measurement and representation of the earth environment (movement of air masses, temperature and precipitation gauges, the erosional action of running water) in laboratories or on display tables such as in the earth sciences. At the outset, it should be noted that geography equipment and facilities can also enhance learning opportunities in current events, international education, global education, history, and earth sciences. The methodology of geographic instruction interconnects with all these other curriculum requirements.

Making maps is an important geography learning activity. This involves learning basic cartographic skills for constructing simple maps to represent earth data. Because many classrooms are equipped with the extended armchair desks, mapmaking activities may become cumbersome or even curtailed. In addition, it makes use of oversized maps for students awkward.

Geography materials, such as maps, globes, and models, require adequate storage space. There are some fairly inexpensive ways to store these materials and there are also some suggested alternatives to conventional wall maps that can reduce the amount of storage space needed. Nevertheless, any well-equipped classroom for teaching geography will have special storage requirements. For example, some large roller maps can remain on special wall fixtures; others may have self-contained stands, but it is advisable to have shelves or map drawers for storing globes, topographic maps, and other teaching materials. Many forms of map storage facilities are possible and some can be constructed inexpensively (Chatham and Vanderford 1969). Many map companies sell storage components for maps, globes, and other geographical materials. Your school budget will determine whether you may purchase these.

PLANNING GEOGRAPHY CLASSROOM FACILITIES

Most schools starting geography programs will not be able to accomplish all that is required for proper geography teaching facilities in one year. With judicious planning and weighing of suitable alternatives, one can prepare the way for phasing in space and facility acquisitions.

The educational environment for a geography program should encourage students to extend their visions from classroom instruction to the world out-of-doors (Boehm and Kracht 1974). It should also encourage open-ended learning and pursue independent thinking beyond course requirements. Fieldwork or field trips must be an integral part of every geography course. Fieldwork need not be elaborate, complex, or conducted at great distances from the school. Many creative observational and data-collecting activities can be carried on within the immediate school environment.

CLASSROOM FACILITIES

There is very little recent literature that deals with the physical facilities required for teaching geography in the classroom. On the other hand, three publications provide excellent guidelines for demonstrating both the criteria required for effective teaching and learning geography and for designing the actual facilities. The most comprehensive of these is by Robert Stoddard (1973). Although planned to deal specifically with college facilities, his work has value in discussing criteria for creating an effective learning environment and also provides methods for evaluating different kinds of classroom facilities.

The second of these, by Mamie L. Anderzhon and John M. Riley (1967), in discussing equipment for a geography laboratory deals directly with establishing criteria for selecting materials, suggests classroom interior designs, and also lists equipment for the ideal geography classroom. A third useful reference, but much more concise, is Warman's checklist for a geography classroom (1969). Despite their publication dates, the general principles in these books and articles still obtain and can easily be updated.

VARIABLE CLASSROOM REQUIREMENTS

Geography classroom requirements will vary considerably from grade level to grade level. In elementary schools where most instruction for several subjects takes place in one classroom, geography facilities will of necessity be minimal but, if carefully planned, can be effective. If the school is large enough to have several sections of the same grade, geography instruction can be staggered throughout the day so that several teachers can share the same equipment in one specially equipped classroom. In middle or junior high schools where special-purpose classrooms may be more common than in K–6 configurations, geography facilities can be concentrated in one or two specially equipped classrooms, depending upon the size of the school.

GEOGRAPHY EQUIPMENT K–6

The following are suggestions for K–6 classroom equipment for instruction in geography. Not all of the suggested equipment is essential; it represents a compromise rather between the minimal and the ideal.

Desks. Most elementary school classrooms have table-type desks with dimensions of approximately 24 by 30 inches. These are generally adequate for spreading out maps and using atlases. At least one or two tracing tables (light tables), 30 inches wide and 5 or 6 feet long should be available. At either end, a frosted glass top with dimensions of 16 inches by 21 inches can be used for individual study and for the use of transparency and overlay maps. These desks can also double as work tables.

Work tables. A large work surface—6 feet by 2½ feet and 30 inches high—should be available. It should have storage space below for topographic, weather, or other large flat maps. Two tables are preferable to one. The teacher can also use this table for demonstration purposes.

Map rails. Maps, pictures, charts, and graphs require considerable wall space. Map rails with movable hooks and clips should be mounted on at least two wall areas (front and nonwindow sides of the room). In schools with open classroom formats, other arrangements will be required.

Ideally, map rails should be mounted at two or three different levels from the floor to accommodate maps of varying lengths. The different levels also permit teachers to use several maps of different sizes, areas, or themes for comparative purposes at the same time. The rails can also be used for storing roller-type maps. Wall maps should be large enough that students can see some details from all parts of the classroom. Readable wall maps should be at least 5½ feet by 4½ feet. Smaller wall maps might be used on the side walls of a rectangular classroom to maximize viewing for all students.

Many wall maps are available in a foldable mounting and have eyelets for wall mounting. These are usually about 20 percent less expensive than the spring-roller mounted maps and have the advantage of requiring minimal storage space—that is, in an ordinary filing cabinet or on a shelf. Some teachers feel that using these maps requires too much preplanning (especially if several classrooms share the maps and their storage is at a place other than in the classroom). Having a series of essential political/physical spring-roller maps mounted in each classroom increases the spontaneity of map use.

Bulletin boards, chalkboards. Most classrooms generally have chalkboards at the front of the room as well as bulletin board space. The ideal geography classroom should have chalkboards on at least one side of the room as well as in the front to provide ample space for drawing diagrams, sketches, outlines, and outline maps. Teachers sometimes have a tendency to use bulletin boards for permanent displays, but the best use for bulletin boards is to have changing displays of student work as well as map, graph, and photograph essays of the region or topic under study in the particular class. Setting aside a portion of the bulletin board space for geography projects will call attention to its importance in the curriculum.

Figure 1
LABORATORY-CLASSROOM FOR GEOGRAPHY FOR BOTH COOPERATIVE AND INDEPENDENT STUDY (ANDERZHON AND RILEY 1967)

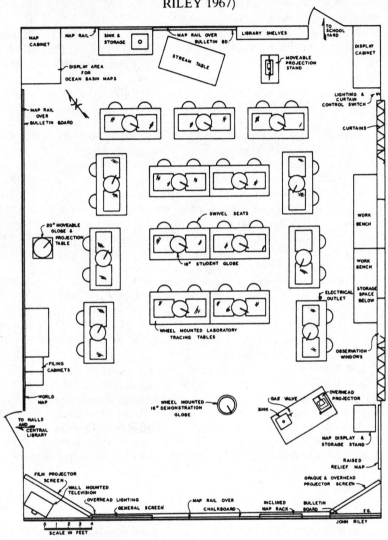

Anderzhon and Riley (1967) provide an exhaustive list and description of essential features for a geography laboratory ranging from optimal room dimensions, storage and display equipment, and classroom equipment (Figure 1). They also make some important observations about ways to improve geographic instruction. They recommend that every two students have a 16 inch physical/political globe and an unabridged world atlas. Teachers should have a demonstration desk with water connections so that students can observe land-form development and stream erosion. Atlases are to studying geography as dictionaries are to the language arts. Students should be encouraged to use the atlas as an essential reference tool not only in geography but in history, language arts, or wherever place names are mentioned.

In addition, Anderzhon and Riley recommend a stream table with a water circulation system for demonstrating student projects in physical geography, i.e., erosional effects of water on soil, delta formation, and stream pattern development.

Warman (1969) provides a handy checklist of equipment and facilities for a geography room. His article contains a schematic drawing of a model classroom (Figure 2).

Map companies and scientific materials suppliers can be helpful in providing an array of materials available for teaching geography. These companies use authoritative educational consultants to develop their products. Their catalogs can also be instructive. Teachers should have well-planned lists of priorities for materials they need in order to spend their limited resources effectively.

Developing an effective geography program can be a tedious and time-consuming process. It requires careful planning to combine the elements of the curriculum, space, equipment, and facilities, as well as the overall objectives of the school's program. But the rewards will benefit hundreds of students by improving their knowledge of the world.

No geography program can be effective unless it relates the information, concepts, and ideas obtained from classroom activities to the real world. The field is the ultimate geography laboratory. The teacher should make every effort to provide students with experience in field study. Gritzner and Larimore (1967) provide an excellent brief summary of how to prepare for and conduct field trips. They point out the importance of teacher planning of areas to be observed and features to be pointed out. They emphasize pretrip briefing ("what you will see and why") and provide a map or maps of the areas to be traversed, work sheets for recording observations, and careful questioning and narration during the trip. A post-trip summary, perhaps with individual assignments, culminates this activity. They point out that it is also safe to assume that students do not know as much about the local area as they think they do. The authors then relate some helpful activities that can serve specific types of field trips.

GEOGRAPHY EQUIPMENT 7–12

Equipment and facilities requirements for geography programs in grades 7–12 are similar to those suggested for grades K–6 but with suitable accommodations for the age differences of the students. Teachers at secondary school levels might argue for special classroom facilities more effectively than those at lower levels because geography frequently has a separate course identity at this level. In some schools, geography may also include considerably more work in the physical and earth sciences than in the lower grades. These courses would demand specialized laboratory facilities.

Figure 2
THE GEOGRAPHY ROOM (WARMAN 1969).

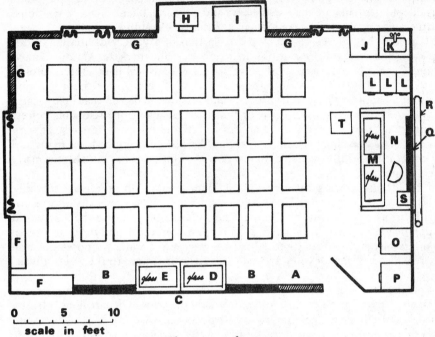

The geography room
Legend

A. Bulletin board
B. Chalkboards
C. Whitewall (can serve as screen)
D. Exhibition cases; map storage
E. Exhibition cases; map storage
F. Bookcases
G. Bulletin board
H. Globe and equipment display stand
I. Storage cabinets
J. Slate slab
K. Sink

L. Filing cabinets
M. Demonstration table with light tables
N. Demonstration surface
O. Map cupboard for sheet maps
P. Cupboard
Q. Chalkboard with wood border for map mounting
R. Auxiliary projection screen
S. Television set and VCR
T. Overhead projector

The regional and topical course descriptions in the *Guidelines for Geographic Education* (Committee on Geographic Education 1984) indicate the need for various subregional and regional maps in these courses. Most map companies publish these maps. The references in the resources section of this chapter will indicate their suppliers. Introducing students to a wide variety of thematic maps beyond the standard physical/political maps is essential at the secondary school level. Teachers will want to have world maps showing the distribution of a variety of cultural phenomena (e.g., population density and distribution, agriculture, trade and transportation, water resources, languages, religions, political ideologies, and political alliances) and physical phenomena (e.g., soils, temperature, rainfall, flora and fauna, and climatic regions).

Teachers should seek to develop a collection of a variety of atlases—local, regional, and thematic. Geographical dictionaries, almanacs, the various United States Census publications, and other standard statistical data sources should also be available or easily accessible in the school library. For those teaching geography courses that emphasize the state, teachers will find that many states have excellent state atlases, most of which are available at reasonable prices. Many states may even offer the atlases to schools for classroom use at very nominal fees. Teachers should not ignore the great variety of local maps and atlases that are readily available for not only the United States but also for many other countries.

United States Geological Survey (USGS) topographic sheets are rich teaching tools for local geography. They can also give students a sense of the environments at other places in the United States. These maps bring the real world somewhat closer than the smaller-scale maps of the region and the state. Aerial photographs and remotely sensed imagery serve the same purposes but also require that students develop special skills for using them to their best advantage.

For physical geography classes, teachers may wish to develop, in phases, a weather-reporting station in cooperation with the science department. Nearby college and university geography departments can also provide teachers with valuable assistance in setting up equipment for physical geography programs.

Teachers who wish their students to develop some proficiency in preparing maps should first attempt to make cooperative arrangements to use the facilities of the mechanical drawing or drafting rooms in their schools before planning to invest large sums for cartographic equipment. Local geography departments in colleges and universities may also be able to provide sound advice for developing a cartography program for grades 7–12.

Computer terminals are common in many junior and senior high schools. Instructional computer programs in geography are being developed at a rapid pace. Teachers should consult the references cited under the resources section in this chapter that deal with instructional programs. Many school systems have instructional media departments that can also provide assistance in obtaining appropriate programs for geography classrooms.

Field trips should be integral parts of geography programs in the secondary schools. The references cited on field trips earlier in this chapter provide some basic guidelines for planning and taking these trips. The skills of asking geographic questions and acquiring geographic information and their explication as noted in the *Guidelines for Geographic Education* and in chapter 7 of this bulletin are useful for developing serious instructional objectives for field trips.

The resourceful geography teacher at both the elementary and secondary school levels should examine the varied resource people and materials available in their schools that might be suitable for their geography programs. Teachers should not ignore the wealth of community resources available for school use, usually available at a minimal cost. Of greatest importance is the need to demonstrate the importance of the geography program to the school and community. Teachers should make their resources known and available as widely as possible to all other programs in the school.

ADDITIONAL HELP

During the past five years, a number of organizations have developed communications networks to assist teachers in improving their teaching effectiveness. Two organizations are ready to assist teachers and schools in improving geography programs. The Geographic Education National Implementation Project (GENIP) has a network of more than 600 teachers and university professors who will assist schools and teachers with their geography programs. GENIP also has developed materials for curriculum developers, principals, and teachers. GENIP is a joint project of four geographical organizations in the United States—the American Geographical Society, the Association of American Geographers, the National Council for Geographic Education, and the National Geographic Society. The last named has also developed a series of state geographic alliances that offer institutes and workshops for teachers and also provide materials and assistance for geography programs in the schools.

REFERENCES

Anderzhon, Mamie L., and John M. Riley. "Equipment for a Geography Laboratory." *Bulletin of the National Association of Secondary-School Principals* 50, no. 316 (February 1967): 53–62.

Boehm, Richard G., and James B. Kracht. "Geographical Expeditions: Field Work for the Elementary Grades." *Journal of Geography* 73 (September 1974): 7–12.

Chatham, Ronald L., and Jay B. Vanderford. "The Wall Map Storage Problem." *Journal of Geography* 68 (February 1969): 93–95.

Committee on Geographic Education. National Council for Geographic Education (NCGE) and the Association of American Geographers (AAG). *Guidelines for Geographic Education: Elementary and Secondary Schools.* Washington, D.C., and Macomb, Ill.: AAG and NCGE 1984.

Eliot-Hurst, Michael E. "Educational Environments: The Use of Media in the Classroom." *Journal of Geography* 72 (November 1973): 41–48.

Geographic Education National Implementation Project, Committee on K–6 Geography. *K–6 Geography: Themes, Key Ideas, and Learning Opportunities.* Washington, D.C.: Geographic Education National Implementation Project, 1987.

Gritzner, Charles F., and Philip B. Larimore. "Media Available within the Local Environment." *Journal of Geography* 66 (May 1967): 9–10.

Stoddard, Robert H. *Planning College Geography Facilities: Guidelines for Space and Equipment.* Commission on College Geography, Publication 12. Washington, D.C.: Association of American Geographers, 1973.

Warman, Henry J. "The Geography Room." *Journal of Geography* 68 (November 1969): 498–99.

PART 3. RESOURCES FOR KEEPING CURRENT IN GEOGRAPHY

Joan Juliette

Today's elementary and secondary teachers have a wealth of information sources to aid in teaching geography skills. Maps, globes, and atlases, the classic tools of the geography curriculum, have been enhanced by the addition of audiovisual presentations, aerial photographs, satellite imagery, computer programs, and geographic information systems. The problem of today's overworked teachers is how to keep up with these new materials. The following ideas suggest some ways of dealing with the problem.

1. **Read the relevant professional journals.** Journals are the best source of current developments, trends, and products in any subject area. Journals on education should be available in every school. These offer ideas for approaches to teaching often developed by other professionals. Social studies skills are often featured. A better source of information is journals dealing with geography or social studies, such as *Journal of Geography* or *Social Education.* If you do not personally subscribe to these journals, request that your school enter a subscription to be shared by several teachers.

   Reviews of textbooks and audiovisual and other teaching materials are often regular features in educational journals. Reviews of computer programs have been recent additions to many journals and help in their selection. Often, producers do not allow the return of computer materials, so it is vital that you base your choices on more than the producer's catalog description. An additional benefit of computer reviews is that programmers can react to criticism by revising the programs to eliminate flaws.

2. **Get involved in a geography or social studies organization.** State and national organizations offer opportunities for professional growth through their publications and local, regional, and national conferences. Local organizations are convenient ways to exchange ideas with others in similar situations as well as review materials on display.

3. **Expand your horizons.** Try using some of the new materials. Have you ever shown your students an aerial photo or satellite image of your area or a videotape of a raft ride down the Colorado River through the Grand Canyon? For students who can't seem to understand direction, have you suggested that they spend some time with a computer tutorial program?

4. **Let your school librarian know of your interest and need for geographic materials.** Your school librarian can be a great source of information on new materials. Librarians routinely use selection guides and reviewing journals, such as *School Library Journal* and *Booklist,* to aid in purchasing materials. Voicing your interest in knowing what is new in geography materials will alert the librarian that you want to keep current

in your field. Offer suggestions of materials you would like purchased for the school library.

5. **Get in touch with your local university's geography department.** The department may be using new techniques or equipment of interest to you. Better yet, take a refresher course or workshop to update your knowledge and skills, or encourage your district to have a geography consultant conduct a workshop during an in-service day.

6. **Become informed about geography networks and alliances.** National attention has been focused on the need for increased emphasis on geography in our schools. The Geographic Education National Implementation Project (GENIP) is an alliance formed by the National Council for Geographic Education, the Association of American Geographers, the American Geographical Society, and the National Geographic Society. A GENIP consultant may be available in your area to offer assistance to educators. In addition, the Geography Education Program at the National Geographic Society can be helpful. Your state may have one of the National Geographic Society sponsored alliances.

The following list of sources for materials and information will help you get started. They are only a sample of the many available to today's educators.

Geographic and Social Studies Organizations and Journals

1. American Geographical Society
   156 5th Avenue
   New York, New York 10010
   (212) 944-2456

The society is devoted to research in geography and dissemination of geographic knowledge. *FOCUS*, one of the publications of the society, provides good background information on various regions of the world in each issue.

2. Association of American Geographers
   1710 Sixteenth Street N.W.
   Washington, D.C. 20009
   (202) 234-1450

AAG is dedicated to furthering "professional investigations in geographic research in education, government, and business." It publishes *Annals, Professional Geographer*, and *AAG Newsletter*. These publications are devoted mainly to advanced geographical research. A useful publication for students and career counselors is *Careers in Geography*.

3. Geographic Education National Implementation Project (GENIP)
   1710 Sixteenth Street N.W.
   Washington, D.C. 20009
   (202) 234-1450

The national headquarters for the GENIP can provide information on publications and programs.

4. National Council for Geographic Education
   Department of Geography and Regional Planning
   Indiana University of Pennsylvania
   Indiana, Pennsylvania 15705

NCGE's purpose is to "encourage the training of teachers in geographic concepts and develop effective geography programs in schools, colleges, and with adult groups." It maintains a goal of stimulating the production and use of geography teaching aids and materials. Its publications include *Journal of Geography, Perspective,* Pacesetter Books, and an instructional activities series. A K–12 teacher membership is available.

5. National Council for the Social Studies
   3501 Newark Street N.W.
   Washington, D.C. 20016
   (202) 966-7840

The national organization of social studies educators publishes *Social Education* and many materials for the social studies curriculum. It promotes and supports social studies education at all levels.

6. National Geographic Society
   17th & M Streets N.W.
   Washington, D.C. 20036
   (202) 828-5699

The society is well known for the *National Geographic Magazine* and its expeditions and research in geography, natural history, astronomy, archaeology, ethnology, and oceanography. It disseminates this knowledge through magazines, maps, books, films, videotapes, educational materials, and television specials. It has recently developed a Geography Education Program.

7. Center for Teaching International Relations
   University of Denver
   Denver, Colorado 80208-0268

   Publisher of activity books.

8. Population Reference Bureau, Inc.
   777 14th Street N.W.
   Washington, D.C. 20005
   (202) 639-8040

   Educator membership available. The bureau produces many useful publications and data bases for geography and social studies classrooms.

9. The World Bank
   1818 H Street N.W.
   Washington, D.C. 20433

   Offers a variety of free or low-cost publications, including teaching units and lessons.

10. State and regional organizations. Get in touch with organizations 1–6 listed for information on how to become members of these groups or their state or regional affiliates.

Journals Reviewing Educational Materials

These are three of the most reputable sources of reviews of books and other educational materials.

1. *Booklist*   Bimonthly
   American Library Association
   50 E. Huron Street
   Chicago, Illinois 60611

2. *School Library Journal*   Monthly
   Bowker Magazine Group
   Subscription Department
   P.O. Box 1427
   Riverton, New Jersey 08077

3. *Curriculum Review*   Bimonthly
   517 South Jefferson
   Chicago, Illinois 60607

Sources of Aerial Photographs and Satellite Imagery

1. Earth Observation Satellite Co.
   4300 Forbes Boulevard
   Lanham, Maryland 20706
   (800) 367-2801

The Landsat products and services available were produced from images taken by the Landsat satellite.

2. U.S. Geological Survey
   EROS Data Center
   Sioux Falls, South Dakota 57198

The EROS Data Center is the national source of aerial photos. The purchaser can obtain a list of available photos on a desired area by filing an inquiry form.

3. Teacher information packets are available from:

   U.S. Geological Survey
   Geologic Inquiries Group
   Mail Stop 907
   National Center
   Reston, Virginia 22092

   NASA Education Offices
   National Headquarters
   Director, Education Program Division
   Code FE
   National Aeronautics and Space Administration
   Washington, D.C. 20546

4. Get in touch with your region's university geography department for custom-designed materials.

Map and Globe Sources

1. ADC
   6440 General Green Way
   Alexandria, Virginia 22312
   (703) 750-0510

Specializing in street atlases of Virginia, Maryland, Washington, D.C., Southeastern Pennsylvania, and Atlanta, Georgia. Features include place names, places of worship, hospitals, block numbers, airports, postal zip codes, schools, parks, and recreational areas.

2. American Map Corporation
   Hagstrom Map Company
   46-35 54th Road
   Maspeth, New York 11378
   (718) 784-0055

Business and education maps, Colorprint and Cleartype brand maps and student atlases, some publications in Spanish, travel maps, lamination and mounting services.

3. American Automobile Association
   8111 Gatehouse Road
   Falls Church, Virginia 22047-0001
   (703) AAA-6000

Publications available to members through local offices.

4. Arrow Publishing Company, Inc.
   1020 Turnpike Street
   Canton, Massachusetts 02021
   (800) 343-7500

Street and road maps, street guides, atlases, mostly East Coast states (Maine, Massachusetts, New Hampshire, Connecticut, Rhode Island, Virginia, Washington, D.C., Maryland, Pennsylvania, Georgia, North Carolina, South Carolina, Florida, New York, New Jersey).

5. Champion Map Corporation
   4237 Raleigh Street
   Charlotte, North Carolina 28213
   (800) 438-7406

Wall maps, map books, folding maps, 1200 cities in 40 states, printed on Canvas Texocloth, lamination and mounting services.

6. George F. Cram, Inc.
   P.O. Box 426
   Indianapolis, Indiana 46206
   (317) 635-5564

Supplier of educational maps, globes, and atlases, as well as audiovisual and computer materials. Distributor for Justus Perthes Map Publishing Company of West Germany.

7. Earth Observation Satellite Company
    4300 Forbes Boulevard
    Lanham, Maryland 20706
    (800) 367-2801

Landsat products and services.

8. Geoscience Resources
    2990 Anthony Road
    Burlington, North Carolina 27215
    (800) 742-2677

Geology, topographic, tectonic, mineral, magnetic anomaly, Bouguer gravity, and relief maps; charts, the Drift Globe, Puzzle of the Plates.

9. Hammond, Inc.
    515 Valley Street
    Maplewood, New Jersey 07040
    (800) 526-4953

Scan globes (illuminated), wall maps, reference maps, student project and bulletin board maps, atlases.

10. Hubbard Scientific Company
     P.O. Box 104
     Northbrook, Illinois 60062-9976
     (800) 323-8368

Educational maps and globes; including Student Landform Series, relief maps, chalk-markable globe, Land Mass Relief Globe with Cross Section of the Earth, Land and Ocean Globe.

11. Michelin Guides and Maps
     Post Office Box 3305
     Spartanburg, South Carolina 29304-3305
     (803) 599-0850

Area travel guides to Europe, including road maps with information on hotels, restaurants, camping sites, tourist attractions, and cultural sights.

12. Murray Hudson Antiquarian Books and Maps
     Route 1, Box 362
     Dyersburg, Tennessee 38024
     (901) 285-0666

Source of antique maps and atlases.

13. The National Survey
    Chester, Vermont 05143
    (802) 875-2121

USGS topographic maps, raised relief, wall, county, U.S., and world maps; globes, map puzzles, antique county atlases, old highway maps, school outline maps.

14. Nystrom
    Division of Herff Jones, Inc.
    3333 Elston Avenue
    Chicago, Illinois 60618
    (800) 821-8086

Atlases, globes, and maps, including relief, historical, state, world, desk, and outline.

15. Erwin Raisz
    130 Charles St.
    Boston, Massachusetts 02114
    (617) 523-4520

Minutely detailed landform maps based on air photographs; time chart of History of Maps.

16. Rand McNally & Co.
    Educational Publishing Division
    Box 7600
    Chicago, Illinois 60680
    (800) 323-1887

Educational maps, globes, and atlases, plastic map sheets; includes Denoyer-Geppert products.

17. Real Estate Data, Inc.
    2398 N.W. 119th Street
    Miami, Florida 33169
    (800) 327-1085

Publisher of information on ownership and physical characteristics of properties for the real estate industry, business community, and government agencies.

18. Sanborn Map Company, Inc.
    629 5th Avenue
    Pelham, New York 10803
    (914) 738-1649

Detailed city maps of more than one hundred cities, including building information, zoning classifications, and land use inventory.

19. Thomas Bros. Maps
    17731 Cowan
    Irvine, California 92714-6065
    (714) 863-1984

Specializing in California, Arizona, Oregon, and Washington; road atlases, street guides, wall maps.

20. Time Education Program
    10 N. Main Street
    Yardley, Pennsylvania 19067
    (800) 523-8727
    (800) 637-8509 in Pennsylvania

Map Pack—wall chart maps of eight regions and continents; Using Maps— filmstrip/cassette on *Time* cartographer, Paul Pugliese. Includes how the projection is chosen, the use of symbols, how geographic data are translated into cartographic display, and follow-up activities on map interpretation.

21. U.S. Geological Survey
    EROS Data Center
    Sioux Falls, South Dakota 57198

Source of aerial photographs from EROS Data Center; inquiry form used to request a geographic search of aerial photos available on the inquiry area.

22. U.S. Geological Survey
    Geologic Inquiries Group
    Mail Stop 907
    National Center
    Reston, Virginia 22092

Source of all existing U.S.G.S. topographic, geologic, and general maps. Write for information and catalog.

23. NASA Education Offices
    National Headquarters
    Director Educational Program Division
    Code FE
    National Aeronautics and Space Administration
    Washington, D.C., 20546

Teachers information packet available.

24. Wilderness Press
    2440 Bancroft Way
    Berkeley, California 94704
    (415) 843-8080

Topographic maps produced by the press on western park areas such as Yosemite and Lassen National Parks.

N.B. One of the most comprehensive sources on maps and charts is *The Map Catalog*, Joel Makower, Editor, and Laura Bergheim, Associate Editor. New York: Vintage Books, Random House, 1986. In addition to providing an introduction to mapping, the book lists a variety of land maps, sky maps, water maps, and map products (atlases, software, globes, map aids, and relief maps), the appendices provides information on where to obtain these maps from federal, state, and international map agencies, selected map libraries, and map stores.

Computer Software Bibliographies

Chartrand, Marilyn J., and Constance D. Williams, eds. *Educational Software Directory, A Subject Guide to Microcomputer Software.* Littleton, Colorado: Libraries Unlimited, 1982.

Epie Institute. *Educational Software Selector,* 2d ed. New York: Columbia University, Teachers College Press, 1985.

Hively's Choice: A Curriculum Guide to Outstanding Educational Microcomputer Programs for Preschool through Grade 9—School Year 1983–84. Elizabethtown, Pennsylvania: Continental Press, 1983.

Minnesota Educational Computer Consortium. *MECC Instructional Computing Catalog.* St. Paul, MECC. (A free catalog of programs available from the consortium.)

1985 Educational Software Preview Guide. International Council for Computers in Education.

O'Connor, Daniel F. *Micros and Social Science: Lesson Plans, A Directory of Software for Achieving Specific Learning Objectives and Procedures for Evaluating Software,* 2d ed., vol. 2. Holmes Beach, Florida: Learning Publications, 1986.

Stanton, Jeffrey, ed. *The Book of Atari Software 1985.* Los Angeles: The Book Company, 1985.

Stanton, Jeffrey, 7th ed. *The Book of Apple Software 1986.* Los Angeles: The Book Company, 1986.

Swift's Educational Software Directory: Apple II Edition. Austin: Sterling Swift Publishing Co., 1984.

Truett, Carol, and Lori Gillespie. *Choosing Educational Software—A Buyer's Guide.* Littleton, Colorado: Libraries Unlimited, 1984. (A basic guide for teachers on how to choose software; lists sources to use in evaluating programs.)

Wang, Anastasia C. *Index to Computer-Based Learning,* 3 vols. Entelek, 1981.

Periodicals on Microcomputing

Classroom Computer Learning   Monthly
Peter Li, Inc.
2451 East River Road
Dayton, Ohio 45439

Computers in the Schools   Quarterly
Haworth Press Subscription Department
75 Griswold Street
Binghamton, New York 13904

Computing Teacher   9/Year
International Council for Computers in Education
University of Oregon
1787 Agate Street
Eugene, Oregon 97403

Creative Computing   Bimonthly
P.O. Box 5213
Boulder, Colorado 80322

Educational Computer Magazine   Bimonthly
Edcomp, Inc.
Box 535
Cupertino, California 95015

Electronic Learning   8/Year
P.O. Box 644
Lyndhurst, New Jersey 07071-9985

Instructor Magazine Computer Magazine for Schools   Annual
Instructor Publications
East 1st Street
Duluth, Minnesota 55802

School Microcomputing Bulletin   Monthly
Learning Publications, Inc.
Box 1326
Holmes Beach, Florida 33509

Audiovisual Materials

1. *Educators Guide to Free Audio and Video Materials*, 33d ed., 1986.
   Educators Guide to Free Films, 46th ed., 1986.
   Educators Guide to Free Filmstrips, 38th ed., 1986.

Classic guides to free materials. User should be cautioned that some items promote products or ideals of sponsoring institution.

Available from:
   Educators' Progress Service, Inc.
   Randolph, Wisconsin 53956

2. *Index to Educational Overhead Transparencies*, 6th ed., 1980.
   Index to Educational Slides, 4th ed., 1980.
   Index to 16mm Educational Films, 8th ed., 1984.
   Index to 35mm Educational Filmstrips, 8th ed., 1985.
   Index to Educational Videotapes, 6th ed., 1985.

Recently developed microcomputer data bases for finding instructional media by subject, title, and key words are now available. NICEM materials are considered the classic source for locating audiovisual materials.

Available from:
   NICEM (National Information Center for Educational Media)
   P.O. Box 40130
   Albuquerque, New Mexico 87196
   Orders: (800) 421-8711
   Information: (505) 265-3591

3. *The Video Source Book*, 7th ed., 1985.
   The National Video Clearinghouse, Inc.
   100 Lafayette Drive
   Syosset, New York 11791

A bibliography of videotapes is currently available.

4. GPN
   Box 80699
   Lincoln, Nebraska 68501

Source of educational slides.

5. Hawkhill Associates, Inc.
   25 East Gilman Street
   Madison, Wisconsin 53703

Filmstrips available on geographic topics

6. Agency for Instructional Technology
   Box A
   Bloomington, Indiana 47402-0210
   (800) 457-4509 or (812) 339-2203

AIT's mission is to strengthen education through technology. It has recently produced, in consultation with GENIP, ten 15-minute video programs and related teacher materials entitled *Global Geography* for middle and junior high school students. Check with your state educational agency on the availability of this important new (1988) program or get in touch with AIT.

Trade Book Bibliographies and Reviewing Sources

Booklist. Chicago: American Library Association, biweekly.
Children's Catalog, 15th ed. New York: H.W. Wilson Co., 1986.
Elementary School Library Collection, 14th ed. Bro–Dart, 1984.
Gillespie, John Thomas, ed. *Best Books for Children, Preschool through the Middle Grades.* 3d ed. New York: R.R. Bowker, 1985.
Junior High School Library Catalog, 4th ed. New York: H.W. Wilson Co., 1980.
School Library Journal. Chicago: American Library Association, monthly.
Subject Guide to Children's Books in Print. New York; R.R. Bowker, 1981.
"Notable Children's Trade Books in the Field of Social Studies," appears in *Social Education* each year, usually in the April/May issue.

Miscellaneous

1. *Access* (eight times a year). Contains current information on materials, events, and news related to global/international education. Many relate specifically to geography K–12. Often includes sample lessons on selected topics.
   The American Forum: Education in a Global Age (formerly, Global Perspectives in Education)
   45 John Street, Suite 1200
   New York, New York 10038
2. *Animal Kingdom* (bimonthly). Articles, features, and photographs about animals in their natural habitat. Reveals much about the world's physical

geography for grade 7 through adult. Book and television reviews are objective and professional.
Animal Kingdom
New York Zoological Society
Bronx, New York 10460

3. *Discover* (monthly). Popular magazine of general interest on topics relating to science and technology. Often carries articles with geographic themes.
Discover
Time and Life Building
Rockefeller Center
New York, New York 10020-1393

4. *History and Social Science Teacher* (quarterly: October, December, March, and May). A Canadian journal of comment and criticism on social education. Each issue contains activities and documents for classroom use. Regularly includes materials useful for upper-level geography teachers.
History and Social Science Teacher
16 Overlea Boulevard
Toronto, Canada M4H1A6

5. *Keeping Up* (quarterly). The news bulletin of the Clearinghouse for Social Studies/Social Science Education. Especially helpful in providing data on materials available through the ERIC/ChESS retrieval system. Teacher can request individualized searches of the ERIC data base for a reasonable fee.
ERIC/ChESS
2805 East Tenth Street
Bloomington, Indiana 47405

6. *Media and Methods* (bimonthly throughout the school year). Evaluates educational products, technologies, and programs. Carries a number of articles on computer usage across the curriculum. Honest and straightforward software reviews, many in geography and related fields.
American Society of Education
1511 Walnut Street
Philadelphia, Pennsylvania 19102

7. *Newsmap* (biweekly during the school year—20 issues). Case studies of world crisis areas with maps. Materials may be photocopied. Each issue contains a teacher's page.
New England Cartographics
East Wilton, Maine 04234

8. *Science and Children* (eight times a year). A journal for elementary and middle school science teachers. Frequent articles on topics of geographic interests. Many include related activities.
National Science Teachers Association
1742 Connecticut Avenue
Washington, D.C. 20009

9. *Science Digest* (monthly) Includes broad coverage of topics related to geography (especially ecology and the habitat). Valuable as a teacher resource.

The Hearst Corporation
P.O. Box 10076
Des Moines, Iowa 50350

10. *The Social Studies* (bimonthly). Articles of general interest to social studies teachers on content, method, and discipline-related issues. Most essays dealing with geography are skill-oriented. Some book reviews.
The Social Studies
4000 Albermarle Street N.W.
Washington, D.C. 20016

11. *World* (monthly). Published by the National Geographic Society. This colorful magazine for elementary school children (3–6) contains articles, activities, and features on a broad range of geographic topics.
National Geographic World
National Geographic Society
Washington, D.C. 20036

12. *World Eagle* (monthly, except July and August). This is a serendipity kind of resource with maps, graphs, and charts. Each issue has a focus section (e.g., USSR, federal budget priorities, etc.). Provides helpful data bases for geography teachers.
World Eagle
54 Washburn Avenue
Wellesley, Massachusetts 02181-9990

13. *World Newsmap* (weekly during the school year—30 issues). Graphic update on world crisis areas with strong map orientation. Includes photographs, charts, political cartoons, plus teaching suggestions and a weekly current events quiz.
Curriculum Innovations
P.O. Box 310
Highwood, Illinois 60040

14. *Zoonooz* (monthly). Generally appealing to middle school students. Contains articles about animal habitats and summaries of research findings. Obvious San Diego Zoo orientation but instructive on geographic topics.
Zoonooz
Zoological Society of San Diego, Inc.
Balboa Park
San Diego, California 92103

CHAPTER 9

TEACHERS ASSESS THE FIVE FUNDAMENTAL THEMES OF GEOGRAPHY

Richard T. Farrell and Joseph M. Cirrincione

In 1984, the National Council for Geographic Education (NCGE) and the Association of American Geographers (AAG) published *Guidelines for Geographic Education: Elementary and Secondary Schools,* a blueprint for revising and revitalizing geography in the public schools (see insert in chapter 1).

The guidelines focus on five broadly defined central themes around which a K–12 curriculum can be developed. These are: location, place, relationships within places (human/environmental interaction), movement, and regions.

The guidelines also outline a sequential framework for organizing a curriculum and suggest some learning outcomes that might be anticipated if such a curriculum were implemented (Committee on Geographic Education 1984).

Clearly the guidelines, particularly the themes, have received wide exposure. The present seems an appropriate time for undertaking a preliminary assessment of how well they have been received by classroom teachers. A questionnaire was mailed to a national stratified random sample of 1,138 social studies teachers (middle school through high school) asking them to rate the importance of each theme on a Likert scale ranging from (1) "not important" to (5) "very important." Responses were received from 594 (52.1 percent). Of these 18 were rejected for various reasons, leaving 576 responses for use in the study. For purposes of analysis, the population was broken down into four main groups—geography teachers, history teachers, civics teachers, and general social studies teachers. Participants were also grouped by the grade level they taught—middle school, junior high school, or senior high school. Teachers self-selected both their primary teaching field and their grade level.

Table 1 summarizes the combined responses of all teachers participating in the study. All four groups of teachers responded favorably to the themes in *Guidelines.* Although there was no significant difference among the teachers regarding the degree of importance of each theme, the rank ordering by mean scores permits some interesting speculation. Those themes traditionally asso-

Table 1. COMBINED TEACHERS' RESPONSES TO THE IMPORTANCE OF
THE *GUIDELINES* THEMES

1. Not Important	4. Generally Important	
2. Marginally Important	5. Very Important	
3. Important		N=576

THEME	Mean	Standard Deviation	Rank
LOCATION	4.00	.95	4
PLACE	4.10	.85	3
RELATIONSHIPS WITHIN PLACES	4.30	.82	2
MOVEMENT	4.33	.81	1
REGIONS	3.96	.92	5

ciated with and basic to geographic instruction—*regions* (a common organizational device for textbooks), *location* (particularly absolute location), and
place (particularly physical characteristics)—were ranked comparatively low.
Movement, with its emphasis on global interdependence, and *relationships
within places,* with emphasis on human-environment relationships, received
higher ratings. Although critics of geography education tend to focus on the
more traditional place and location topics, classroom teachers apparently respond
more favorably to those stressing the global dimension and the environment,
both of which are closely associated with recent developments in social studies
education.

Table 2 shows participants' responses by teaching fields. Although there was
no significant difference in how teachers in different social studies fields viewed
the themes, several observations seem in order. First, geography teachers dis-
played a higher degree of uniformity in rating each theme than other groups.
This can be explained, perhaps, by the fact that geography teachers understand
or at least recognize the interrelationships among the themes more readily than
other teachers. Second, when comparing rank order, the relatively high level
of agreement between geography and history teachers should be noted. This
is somewhat surprising, considering that history teachers seem to prefer taking
a skills/fact orientation toward the role of geography in the curriculum.

The third observation is not as easily explained. Note the unusually high
ratings that civics and general social studies teachers gave to movement and
relationships within places and the comparatively low ratings (particularly
among civics teachers) given the other themes. Perhaps they perceive that the
content associated with movement and relationships within places is more
relevant to their fields than the content associated with the other three themes.
As social studies teachers (as opposed to geography teachers), they would give

Table 2. COMPARISON OF MEAN SCORES AND RANKINGS OF SUBJECT TEACHERS' RESPONSES TO THE IMPORTANCE OF THE *GUIDELINES* THEMES

1. Not Important	2. Marginally Important	3. Important	4. Generally Important	5. Very Important	N = 576

	GEOGRAPHY TEACHERS			HISTORY TEACHERS			CIVICS TEACHERS			SOCIAL STUDIES TEACHERS		
THEME	Mean	Standard Deviation	Rank	Mean	Standard Deviation	Rank	Mean	Standard Deviation	Rank	Mean	Standard Deviation	Rank
LOCATION	4.15	.89	(3)	4.03	.95	(4)	3.89	.94	(5)	3.89	.97	(5)
PLACE	4.15	.73	(3)	4.09	.86	(3)	4.09	.81	(3)	4.06	.94	(3)
RELATIONSHIPS WITHIN PLACES	4.25	.86	(2)	4.22	.83	(2)	4.32	.68	(1)	4.41	.84	(2)
MOVEMENT	4.27	.85	(1)	4.27	.79	(1)	4.31	.86	(2)	4.44	.79	(1)
REGIONS	4.04	.92	(5)	3.89	.90	(5)	3.99	.98	(4)	4.02	.94	(4)

greater emphasis to movement and relationships and leave the primary responsibility for the other themes to the geographers.

Table 3 shows the participants' responses by grade level taught. There was no significant difference in the way teachers rated each theme. Although there was general agreement at all levels regarding the importance of the themes, movement and relationships within places again received the highest ratings. More interesting, perhaps, is the rank order of the themes within each group. Place and regions ranked fourth and fifth among middle school teachers, who

Table 3. COMPARISON OF MEAN SCORES AND RANKINGS OF MIDDLE SCHOOL, JUNIOR HIGH SCHOOL, AND HIGH SCHOOL TEACHERS' RESPONSES TO THE IMPORTANCE OF THE *GUIDELINES* THEMES

1. Not Important	2. Marginally Important	3. Important	4. Generally Important	5. Very Important

	MIDDLE SCHOOL TEACHERS			JUNIOR HIGH SCHOOL TEACHERS			HIGH SCHOOL TEACHERS		
THEME	Mean	Standard Deviation	Rank	Mean	Standard Deviation	Rank	Mean	Standard Deviation	Rank
LOCATION	4.05	.83	(3)	4.00	.92	(4)	3.99	.97	(4)
PLACE	4.02	.81	(5)	4.20	.74	(3)	4.08	.85	(3)
RELATIONSHIPS WITHIN PLACES	4.20	.96	(1)	4.37	.75	(1)	4.26	.79	(2)
MOVEMENT	4.16	.88	(2)	4.28	.83	(2)	4.32	.81	(1)
REGIONS	4.03	1.00	(4)	3.80	.90	(5)	3.94	.90	(5)

might be expected to give these themes high ratings because of the content usually associated with them. Similarly, location (fourth) and regions ranked low among junior high school teachers. A cursory review of textbooks used at this level suggests that both themes should have received high ratings.

Conclusions and Observations

This preliminary analysis raises more questions than it answers. The respondents were clearly favorably inclined toward the five themes, rating them between "very important" and "generally important." On the surface, this may appear encouraging, but the apparent lack of discrimination in the ratings of the themes may create problems for curriculum developers. This favorable but indiscriminate response could indicate a critical dilemma facing social studies teachers. They see the need to expand the role of geography in the curriculum but lack the training to identify clear areas of need and to define what that role should be.

There is also the possibility that their responses reflect what they teach in the classroom rather than a strong commitment to the themes. The high ratings given movement and relationships within places, both current and popular topics in social studies, and the comparatively lower ratings given the more traditional themes of regions and location may support this generalization.

On the positive side, the high level of agreement on the importance of themes may reflect the participants' awareness of the interrelatedness of the themes. An understanding of one theme facilitates understanding of others; deciding whether one is more important than another is not relevant. If this observation is correct, it suggests a higher degree of sophistication than expected.

One thing is clear: social studies teachers are currently predisposed toward increasing and improving the status of geography in the curriculum. Generating enthusiasm for reform seems less important than the need to help direct change and to provide guidance in curriculum development. Continued elaboration of the themes is necessary if they are to be employed as the framework for identifying specific goals and objectives for the program. The development of a K–6 geography volume and an anticipated volume on 7–12 geography are steps in this direction (GENIP K–6 Committee 1987).

REFERENCES

Committee on Geographic Education, National Council for Geographic Education, and Association of American Geographers. *Guidelines for Geographic Education: Elementary and Secondary Schools.* Washington, D.C., and Macomb, Ill.: AAG and NCGE, 1984.

Geographic Education National Implementation Project (GENIP). Committee on K–6 Geography. *K–6 Geography: Themes, Key Ideas, and Learning Opportunities.* Macomb, Ill.: National Council for Geographic Education, 1987.

AUTHOR INDEX

A

American Geographical Society (AGS), x, 107
American Historical Association, 26
Anderson, Charlotte C., 73
Anderzhon, Mamie L., 99, 101
Association of American Geographers (AAG), ix, 24, 107

B

Bacon, Phillip, 73
Barnes, Charles C., 25
Barr, Robert D., 29, 30, 32
Barrows, Harlan H., 25
Barth, James C., 29, 30, 32
Barton, Thomas, 73
Bengston, Nels A., 25
Berg, Harry D., 73
Boehm, Richard G., 99
Bowman, Isaiah, 26
Boyles, B., 22
Bragaw, Donald M., 42
Butts, R., 26

C

Carpenter, Helen McCracken, 73
Carter, Jimmy, ix
Chamberlain, James, 24, 25
Chatham, Ronald L., 98
Cirrincione, Joseph M., 2
Cobb, Russell, 73
Commission on Foreign Language and
    International Studies, ix
Cornbleth, Catherine, 74
Council on Learning, ix
Cremin, Lawrence, 26

D

Davis, William Morris, 23, 24

E

Educational Testing Service (ETS), ix
Edwards, Mike, 63
Eliot, T.S., 1
Eliot-Hurst, Michael E., 98

Engle, Shirley, 29
Eratosthenes, 1

F

Fair, Jean, 32
Fairbanks, Harold W., 23
Farrell, Richard T., 2
Fenton, Edward, 58

G

Gale, Robert Peter, 59
Garrison, William, 6
Geographic Education National
    Implementation Project (GENIP), ix, 57,
    107
Goodlad, John, 23
Gould, Peter, 6
Greenwald, John, 65, 66
Gritzner, Charles F., 9, 102
Groisser, Philip, 74
Grosvenor, Gilbert M., 51
Guyot, Arnold, 23

H

Hanna, Paul, 26
Hart, John Fraser, 6
Hill, A. David, ix
Hirsch, E.D., Jr., 2
Holmes Group, 52
Houston, Edwin, 24
Hudson, Ray, 6

J

James, Preston, 23, 24, 25, 26, 73
Joyce, Bruce, 74
Joyce, W., 73

K

Kemball, Walter G., 52
Kilpatrick, James, 75
Kracht, James B., 99
Kurfman, Dana, 29

L

Larimore, Philip B., 102
Lemaire, Minnie E., 27

M

Manson, Gary, 54, 73
Martin, Geoffrey J., 24
Maryland State Department of Education, 18
Mayo, William L., 27, 28
Mehlinger, Howard, 28
Morrill, Richard, 3
Morse, Jedidiah, 22

N

Natoli, Salvatore J., 1, 6, 11, 28
National Council for Geographic Education (NCGE), ix, 97, 108
National Council for the Social Studies (NCSS), 11, 27, 73, 97, 108
National Education Association, Committee of Ten, 24
National Geographic Society (NGS), x, 108
National Science Teachers Association, 57

P

Pattison, William D., 33
Pestalozzi, Johann, 23
Piaget, Jean, 27

R

Ramberg, Bennett, 66, 67
Rice, Marion, 73
Riley, John M., 99, 101
Ritter, Karl, 23
Robinson, Arthur, 86
Rosen, Sidney, 23
Rumble, Heber, 22, 23

S

Share, H., 23
Shermis, Samuel S., 29, 30, 32
Southern Governors Conference, 11
Stoddard, Robert, 99
Stoltman, Joseph, 23, 27
Stout, John E., 23
Stowers, Dewey, 23, 24

V

Vanderford, Jay B., 98
Vobejda, Barbara, 11
Vuicich, George, 23, 54, 73

W

Warman, Henry J., 99, 102
Warntz, William, 22, 23
Weil, A., 27, 74
Winston, Barbara, 11, 28, 55, 73
Wise, John H., 59

SUBJECT INDEX

A

Age-mate geography. *See* Curriculum, age-mate geography

American Geographical Society (AGS), x, 107

Association of American Geographers (AAG), ix, x, 24, 107, 119

C

Cartography, 7, 86; computer, 7

Census, U.S., as a source of geographic data, 77–78; computer data bases, 90

Certification, teacher. *See* Teacher certification

Chernobyl nuclear accident, 2, 59ff.; teaching activities, framework for, 61–68

Citizenship, 9, 29, 36

Classroom facilities, planning, 99ff. *See also* Geographic education, equipment and space

Computers and microcomputers, 7, 89, 90, 97; periodicals about, 114–115; software, 90, 97; software bibliographies, 114

Critical thinking, 39

Cultural geography. *See* Geography, cultural

Cultural Literacy, E.D. Hirsch, Jr., 2

Cultural literacy, 2, 8

Current/Contemporary events, teaching/learning activities, 59, 70

Curriculum, age-mate geography, 27; assessment of, 94, 95; decision making, teachers' role in, 33, 94–97; elementary, 22, 23, 26, 27, 42; expanding horizons model, *see also* "expanding horizons" model; incorporation of geography into, 8, 11, 20, 94–97; journey geography, 27; reform movements, 11, 28; science, 25; secondary, geographic skills for, 74; social studies, 15

Curriculum development, 1, 16, 27, 119, 122; course proposals and grant applications, 95; *see also* Education reform

D

Data bases, 89, 90

Decision making, 29, 31; development of, 73

Development, economic, 6, 89

Development, psychological, 26. *See also* Curriculum, spatial competence

Distance-decay functions, use of, 79 (table), 80 (figs.)

E

Economic geography. *See* Geography, economic

Education reform, ix, 11, 24, 26

Environmental education, 28

Expanding horizons model in the geography curriculum, 26, 33, 43ff.; awareness of self; self in space, 43; individual in primary social groups; homes and schools in different places, 43–44; meeting basic needs in nearby social groups; neighborhoods, small places and larger communities, 44–45; the community; sharing space with others, 45–46; human life in varied environments; the region, 46–47; people of the Americas; the U.S. and its neighbors, 47; people and cultures; the Eastern hemisphere, 48–49; a global view; state, regional, or world geography, 49–50; *see also* Geographic education, scope and sequence for

F

Fieldwork/Field trips, 76–77, 78, 99, 104

Focus vs. coverage, 36

Fundamental themes of geography, illustrated, 4–5; listed, 3, 33, 39, 52, 53, 61, 69–70, 73, 119; teachers' assessment of, 119ff.; use in scope and sequence, 43–50, 76; location; absolute and relative, 2, 43–49, 61, 62, 76; movement (spatial interaction), 43–50, 65–67; place, 43–49, 62, 63; regions, 6, 43–50, 67, 68; relationships within places (human/environmental relationships), 43–49, 63, 64

G

Geographic education, content of, 20 (table); contribution to the social sciences tradition, 32–33; elementary schools, x, 23; equipment and space, 101 (fig.), 103 (fig.); evaluation of, 73; graduation requirements, 11; history of, 22ff.; learning opportunities, 42ff.; model geography laboratory, 101 (fig.)–102; objectives, 1, 19 (table), 72; relevance, 36; scope and sequence for, 26, 73, *see also* "expanding horizons" model; secondary schools, x, 23, 27; statement of skills for, 74; teacher attitudes, 15, 16 (table)

Geographic Education National Implementation Project (GENIP), x, 51, 57, 105, 107

Geographic information, *see also* Geography, skills; definition of, 86; interpretation of, 78–82, 79 (table); presentation of maps, 25, 82, 83–84 (fig.); presentation of tables and graphs, 78, 79 (fig.), 80 (figs.); sources of, 76–78, 89–90; written and verbal skills, 85

Geographic Information Systems (GIS), 7

Geographical literacy, ix, 2, 8; role of curriculum planning, 97

Geography, careers in, 7; cultural, 13, 53–56; definitions of, 1, 3, 9, 28; economic/ commercial, 13, 25, 27, 55, 56; geomorphology, 55, 57; historical, 54, 55, as preparation for teaching history, 55; physical, 23, 24, 27, 53–57; political, 56; population, 56; rationale for teaching as a separate subject, 7; regional, 6, 13, 24, 53, Canada, 55–56, developing world, 55, North America, United States, 54–56, research status of, 6, South America, 55, Western Hemisphere, 55; skills, 34, description of fundamental, 72ff.; systematic, 15; traditions of, 13, 22, 26, 33, 39; urban, 56; world/global, 13, 53–56, developing and testing generalizations, 87–89, teacher preparation in, 15; *see also* Teacher preparation

Geography Education Program of the National Geographic Society, 107, 108

Geography in the social studies, as a perspective, 33, 38–40, 38 (fig.); as a social science, studying spatial organization, 36, 37 (fig.); as a topic, studying places, 34–36, 35 (fig.)

Geography networks and alliances, 105, 107

Global education, 28, 97

Guidelines for Geographic Education, ix, 33, 43, importance to teachers, 3, 28, 59, 119, 120 (tables), 121; statement of major geographic skills, 74

H

High School Geography Project (HSGP), 28

Historical geography. *See* Geography, historical

I

International education, ix, 28

J

Journal of Geography, 97, 106

Journey geography. *See* Curriculum, journey geography

K

K-6 Geography: Themes, Key Ideas, and Learning Opportunities, 42

Kindergarten, 43

L

Learning reinforcement, 36

Libraries, 77, 106–107

Location. *See* Fundamental themes of geography

Location analysis, 7

M

Maps, 3, 23, 27, 98, 100, in presentation and interpretation of geographic information, 82–90; interpreting patterns and distributions, 86; map reading skills, development of, 73; mental maps, 9, 76; projections, 86; sources, 104, 110–113

Media, maps, pictures, texts, 35, 77; sources of, 115–116

Motivation, 36

Movement. *See also* Fundamental themes of geography, spatial interaction

Multicultural education, 29

N

National Council for Geographic Education (NCGE), ix, x, 108, 119

National Council for the Social Studies (NCSS), 27, 30, 73, 108; scope and sequence, 33, 42, 73

National Education Association (NEA), 24

National Geographic Society (NGS), x, 108

National Geography Awareness Week, x

National Science Teachers Association (NSTA), 57

Neighborhood, 44–45

P

Physical geography. *See* Geography, physical

Place. *See also* Fundamental themes of geography; sense of, 72, 77; symbolic representation of, 73

Planning, urban and regional, 6

Political geography. *See* Geography, political

Population geography. *See* Geography,
  population
Problem solving, 7, 32, 40
Progressive Education Movement, 26. *See
  also* Education reform

Q
Questions, asking geographic, 74–76. *See also*
  Geography, skills

R
Reflective inquiry. *See* social studies,
  traditions, reflective inquiry
Regions, 6, 27
Regional geography. *See* Geography, regional
Relationships within places (Human/
  environmental relationships). *See*
  Fundamental themes of geography
Relevance, 34
Remote sensing, 3, 7, aerial photo
  interpretation, 55, 56, 57, 87; sources of
  material, 109-110

S
School-community relations, role in
  curriculum development, 96
Science, 6, teachers, earth science, 55, 57
Scope and sequence. *See* Social studies, scope
  and sequence
Social Education, 97, 106
Social studies, curriculum, x; definition of, 29,
  32; diversity, value of, 32; elementary
  school, 26; scope and sequence, 26, 42ff., of
  map and globe skills, 73; traditions, 22, 29,
  30, 31 (fig.); and geography, 33–36, 35
  (fig.); citizenship transmission, 30, 33–36,
  35 (fig.); developmental aspects of, 32;
  reflective inquiry, 31, 38 (fig.), 39; social
  studies as social science, 30, 36, 37 (fig.)
Space, Spatial, definition of, 82
Spatial competence, 27, 73
Spatial distribution, of malaria, 83–84 (fig.)
Spatial interaction. *See* Relationships between
  places
Spatial organization, geography as, 33

Spatial patterns, 86, 88
Survey of Global Understanding, ix

T
Teacher attitudes, current offerings in
  geography, 16 (table); geographic
  education, 16 (table); organizing geographic
  instruction, 16, 17 (table), 20
Teacher certification, 13, 14, 52, 57
Teacher preparation, x, 8, 11, 13 (table), 28,
  geography 13 (table), 14 (table, fig.), 51–58,
  courses and course sequences, 53–57;
  elementary teachers, 54; essential
  requirements, 51–53; high school earth
  science teachers, 57; high school geography
  teachers, 56–57; high school social studies
  teachers, 55–56; in-service training, 21, 31,
  96; middle/junior high school science
  teachers, 55; middle/junior high school
  social studies teachers, 54–55; minimum
  requirements for all teachers, 53–54; profile
  of academic coursework, 12; qualifications
  for teacher trainers, 54
Teaching/learning activities, Chernobyl
  nuclear accident, 61–69; geography in the
  news, 70; global crises, 69–70; interpreting
  maps, 86–87; interpreting tables and graphs,
  78–82; teaching aids, 97–98
Textbooks, 1, 11, 18, 22, 23, 25, role in map/
  globe skills development, 73; sources for
  reviews of, 106
Themes of geography. *See* Fundamental
  themes of geography
Transportation, 7

U
United States Congress, x
Urban geography. *See* Geography, urban

V
Values, 29, 32, 34, 39, perception and, 77

W
World geography. *See* Geography, world
Writing, 39, 40

Index prepared by Alan Turnbull, University of Maryland